HILLSLOP

T0227702

HILLSLOPE FORM

A.J. PARSONS

Routledge
Taylor & Francis Group

LONDON AND NEW YORK

First published 1988
by Routledge

2 Park Square, Milton Park, Abingdon, Oxon OX14 4RN
711 Third Avenue, New York, NY 10017, USA

First issued in paperback 2016

Transferred to Digital Printing 2007

British Library Cataloguing in Publication Data

Parsons, A.J.
   Hillslope form.
   1. Hills, Slopes. Geophysical processes
   I. Title
   551.4'36

Library of Congress Cataloging-in-Publication Data

Parsons, A.J.
   Hillslope form / A.J. Parsons.
   p. cm.
   Bibliography: p.
   Includes index.
   ISBN 0-415-00905-7
   1. Slopes (Physical geography) I. Title.
   GB448.P37   1988
   551.4'36–dc 19   88-20106
                       CIP

ISBN 0-415-00905-7

Cover design: Herman Yiu

ISBN13: 978-0-4150-0905-8 (hbk)
ISBN13: 978-1-138-99220-7 (pbk)

CONTENTS

TABLES

FIGURES

PREFACE

Geomorphology remains largely a field-based science. Notwithstanding the significant advances during recent years in the theoretical and laboratory-based aspects of the subject, most books in geomorphology continue to reflect the locations in which their authors have worked and studied. I have had the good fortune to examine hillslopes in a variety of climatic and geologic settings. As a research student of Ronnie Savigear, I was given opportunities to travel well beyond my own field area in southern Italy. Projects in Morocco, Australia and Argentina allowed me the chance to see hillslopes in these countries as well as in others that I could find the time to visit on my way to or from fieldwork. My susequent career has taken me to other parts of Europe, Africa and Australasia, and more recently to North America. I am particularly grateful to Ronnie Savigear for the early opportunities and to others who have contributed, one way or another, to enable me to continue learning about hillslopes.

Despite the opportunities, gaps remain. I am aware that the book draws most of its examples from humid temperate and warm semi-arid to arid climates. The former is the environment in which I have, for the most part, lived: the latter is the climatic setting in which I prefer to undertake fieldwork. The book would be improved by a wider range of examples and I apologise in advance to any who read this book and whose interests lie in other environments. To them the book may seem excessively biased.

In recent years, my fieldwork has benefited from the support of my wife, Susan. Many times she has acted as an unpaid, conscripted field assistant. On the occasions of my longer field trips, when she has been unable to accompany me, Susan has without complaint taken on the single-handed running of our household. For these things, and for her help and encouragement in preparing this book, I am profoundly grateful.

Finally, I would like to express my thanks to Muriel Patrick and Don Morris of the Geography Department at Keele. They have shown amazingly good humour in coping with my difficult diagrams and (they tell me) my illegible handwriting.

ACKNOWLEDGEMENTS

Many of the Figures and Tables in this book have been reproduced from or are based on ones in other publications. I am grateful to all those concerned for permission to use these materials. Sources are acknowledged individually in the captions to Figures and Tables.

INTRODUCTION 1

The study of hillslopes is a central element of geomorphology. Many authors (for example Scheidegger, 1961a; Young, 1972; Selby, 1982a; Abrahams, 1986) have pointed to the ubiquity of hillslopes over the earth's land surface to indicate the overall importance of their study, and Chorley (1964a) demonstrated that many of the major methodological disputes in geomorphology have been focused on hillslopes. An example is the debate in the latter part of the nineteenth and early part of the twentieth centuries on how landscapes change through time which essentially reduces to a question of the manner of hillslope evolution. Paradoxically, Carson and Kirkby (1972, p.1) were able to point to the comparative neglect of hillslopes when compared to the level of study accorded to other geomorphic features. These authors identified several factors contributing to this comparative neglect, not least being the very ubiquity of hillslopes which, if ignored, reduces the significance of any study and, if faced, raises massive sampling problems. They argued that geomorphologists have preferred to study simpler features, particularly those on which processes operate to effect more rapid change than is common on many hillslopes.

Like all branches of geomorphology, the study of hillslopes has been affected by fashion. Two major periods of enquiry can be identified, each with its own focus of interest. In the early years, the emphasis was upon landscape history and the long-term evolution of hillslopes. This was the period that gave rise to Davis's notion of long-term decline in hillslope gradient (1899), Penck's concept of slope replacement (1924) and Wood's four-element model (1942), amongst others.

1

If an end to this period can be identified, it is probably marked by the publication of King's proposition of parallel retreat of hillslopes (1953). The period was characterised by conjecture and argument but also by a comparative absence of measurement. The period which has followed may almost be regarded as a mirror image of its predecessor; strong on measurement but often weak in argument. At first, measurement focused on hillslope form. Fair (1947, 1948a,b) and Savigear (1952) were pioneering in establishing this field of hillslope geomorphology. In the early part of this period, during which the influence of the previous fashion remained strong, there was a tendency to use measurement of hillslope form to enhance interpretation of hillslope evolution. Later, however, measurement became an adjunct to more quantitative description of hillslope form (e.g. Parsons, 1976a) as the study of hillslopes was affected by the fashion for morphometry. Well after the beginning of this period, the focus of attention began to move to measurement of processes (e.g. Kirkby, 1967). It is this focus that dominates hillslope geomorphology at the present time. In a review of hillslope studies over the period 1970-1975, Young (1978) demonstrated that process studies formed the largest group of hillslope publications during that time. A similar study for the period 1976-1985 shows this group to be the increasingly dominant aspect of hillslope geomorphology (Table 1).

Concern for the manner of hillslope evolution has not been entirely absent from the latter part of the measurement period. For the period 1970-1975 Young recorded few studies that he classified under the heading of hillslope evolution but the period 1976-1985 shows an astonishing increase in this category. Although direct interpretation of the evolution of individual hillslopes from measurements of form and/or process is now seldom encountered in the literature, modelling of the long-term effects of particular processes continues to attract attention. It is in its contribution to such process-response models that measurements of hillslope processes have contributed towards our understanding of hillslope evolution. From the earliest days of hillslope geomorphology there have been those workers who have approached the problem of hillslope evolution by deducing the landforms that would result from

2

Table 1.1: Classification by subject of publications on hillslopes, 1970-1975

Subject	Publications 1970-75	Publications 1976-85	†Relative change
GENERAL TOPICS, TEXTBOOKS, REVIEWS	16	18	-32.5
METHODS AND TECHNIQUES	81	92	-31.9
of which: Instrumentation of processes	18	13	-54.7
Profile analysis	15	32	+28.0
Mapping	21	17	-51.4
PROCESSES (EXCLUDING LANDSLIDES)	180	370	+23.3
of which: Rates of processes	132	26	-88.2
LANDSLIDES	66	290	+163.6
THEORY	29	135	+179.5
of which: Models	22	81	+120.7
CLIFFS AND SCREES	42	28	-60.0
of which: Screes	28	6	-87.2
ANGLE AND ANGLE FREQUENCY	19	46	+45.1
HILLSLOPE FORM INCLUDING PROFILE FORM	108	102	-43.5
ENVIRONMENTS	95	117	-26.3
of which: Specific climates	34	29	-48.9
Specific rock types	20	7	-79.0
Polar and montane zone	15	27	+8.0
Rain forest zone	11	7	-61.7
Pediments	16	14	-47.6
Microrelief	16	13	-51.3
Regolith	15	9	-64.0
HILLSLOPE EVOLUTION	4	99	+1477.6
APPLIED STUDIES	19	52	+64.2
of which: Engineering applications	12	39	+94.6

† Relative change denotes the change in the number of publications between 1970-75 and 1976-85 as a proportion of the number in the former period (adjusted for the difference in the length of the two periods).

Sources: Data for the period 1970-75 taken from Young (1978, p.76) *Slopes: 1970-75*, Oxford University Press, and reproduced by permission. Those for the decade 1976-85 compiled from entries for this period listed in *Geo Abstracts* 1976-86

the operation of specific processes upon an initial hillslope form. To begin with, such modelling was confined to predicting the evolution of a vertical or near-vertical cliff under the processes of weathering and rockfall (Fisher, 1866). Such evolution is relatively easy to model because it can be assumed that rockfall operates instantaneously to remove weathered material. No such simple assumptions can be made about other processes acting upon hillslopes so that any realistic attempts to model the evolution of initial forms under the operation of such

3

processes as rainsplash and overland flow could not take place until some information on the controls of the rates of sediment removal by these processes became available. With the information provided by measurements of hillslope processes modelling of the evolution of hillslopes under various processes has become possible (e.g. Kirkby, 1971).

Rather less of the information that has come from measurement of hillslope form has been incorporated into modelling. At the present time a model is considered successful if the outcome looks anything like a hillslope. Statistical comparisons of the predictions of process-response models to characteristics of natural hillslopes (e.g. Ahnert, 1970a; Abrahams and Parsons, 1977) have been rare and only a few studies have attempted to use process-models to predict actual hillslope forms (e.g. Parsons, 1976b; Kirkby, 1984). The present failure to link measurements of hillslope form to those of processes, either through process-response models or in some other way, remains a major weakness of hillslope geomorphology.

Notwithstanding the advances that have taken place during the last fifteen to twenty years in modelling the effects of a variety of hillslope processes, geomorphologists are still far from understanding the nature of hillslope evolution which Young (1978) described as the central problem of hillslope geomorphology. Several reasons may be advanced to account for this state of affairs. First, most hillslopes evolve under the operation of several processes. The relative importance of individual processes remains to be established. In particular, the relative importance on any hillslope of sporadic mass movement processes as opposed to the effects of continuous wash and creep processes is unknown. Young (1972, p.86) pointed to the volume of material moved by a single small landslide in comparison to observed rates for continuous processes and concluded that with even very low frequency the sporadic processes are responsible for the major part of denudation. In contrast to this deduction is the evidence of the majority of valley sides which display little or no evidence of the operation of sporadic mass movements. In all probability the relative importance of sporadic and continuous processes changes through time so that it is unrealistic to model hillslope evolution under the assumption that the same processes operate with the same intensity through time. Second, most hillslopes evolve not only in response to the processes operating upon them but also in response to conditions at their bases. Although some attempts have been made to incorporate the effects of different basal conditions (e.g. Bakker and Le Heux, 1952, and more recently Kirkby, 1986) no model has faced the issue of realistic changes in these conditions, particularly important in the case of evolution of valley sides where lateral migration of the basal stream channel will result in ever-varying basal conditions. Third, evolution of hillslopes takes place under changing climatic conditions. Although some evidence of the relative importance of different processes under various climatic conditions is now coming forward (Saunders and Young, 1983), no attempt has yet been made to incorporate this information into models.

There can be no doubt that the measurements made over the last forty years have added substantially to our knowledge of hillslopes. Nevertheless, substantial gaps remain. It is the aim of this book to describe the present state of knowledge of hillslope form, to emphasise the results of the measurements made during this period, and to point to unresolved problems in the understanding of hillslope form.

The early chapters are concerned with attempts to measure and describe hillslope form. The task is a difficult one, made more so by the somewhat vague understanding of what constitutes a hillslope. If the land surface of the earth is composed largely of hillslopes, as seems to be widely held, then, logically, these hillslopes must be definable entities. Regrettably, as Dylik (1968) has pointed out, an adequate definition of these entities is lacking and there is confused terminology. *Slope* and *hillslope* are often used interchangeably to refer to (albeit undefined) units of the ground surface but, in addition, *slope* is often used to refer to the property of inclination. In this book the terminology that will be used is as follows.

1. The three-dimensional entities are referred to as *hillslopes*.
2. The property of inclination is termed *gradient*.
3. The term *slope* will be used adjectivally to describe part of a hillslope e.g. slope

5

Introduction

unit; as a qualified noun in accordance with common usage e.g. talus slope; and as a synonym for gradient.
The problem of the definition of a hillslope is not so easily solved. Even for the simple case of a first-order drainage basin in which the upper and lower boundaries of hillslopes present relatively few problems of definition, Parsons (1982) recognised the difficulty of picking suitable lateral limits. His proposal to regard a first-order drainage basin as being composed of two hillslopes, each bounded laterally on one side by the line of the drainage divide from the interfluve crest to the stream at the outlet of the basin and on the other by the line extending the stream head to the interfluve (Figure 1.1), may be conceptually sound but it is unsatisfactory in morphological terms. Both lateral boundaries, and particularly the latter one, may have weak morphological expression. The significance of the problem of hillslope definition for measurement and description, and the role that these activities may play in solving the problem form part of chapters 2 and 3.

Figure 1.1: Component hillslopes of a drainage basin

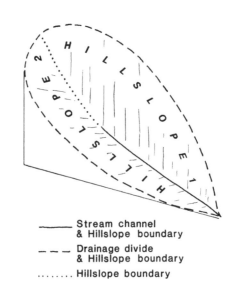

_____ Stream channel
& Hillslope boundary

_ _ _ Drainage divide
& Hillslope boundary

.... Hillslope boundary

6

The remainder of the book deals with the observed variation in hillslope form over the surface of the earth. Chapters 4, 5 and 6 examine the influence of factors that themselves vary spatially over the earth's surface, namely climate, materials and local conditions. Chapter 7 considers the relationship between hillslope forms and processes. At it simplest, this relationship is one of cause and effect but because hillslope processes vary both spatially and temporally this leads to a more complex interaction and consequently a less-than-straight-forward set of relationships. In the long term, hillslope forms result from the operation of processes, but in the short term processes are constrained by existing hillslope forms. Furthermore, the many processes that operate simultaneously on most hillslopes act over different timescales to produce different types of effect on hillslope form. Because hillslope processes are subject to both spatial and temporal variation, chapter 7 forms a link between the three before it, concerned wholly with spatially varying influences on hillslope form, and the two that follow it which examine the relationships between hillslope form and time.

The final chapter considers hillslope form in relation to man. Over much of the earth's surface many hillslopes are now used by man or have been modified by his activities. In addition, man creates artificial hillslopes, for example in quarry faces or road cuttings, and, more extensively, in land reclamation sites may fashion entire landscapes. The implications of these activities and the contribution that understanding of natural hillslope form may make to the management of such anthropogenic hillslopes are examined in this last chapter.

MEASUREMENT 2

Measurement of hillslope form can be defined, in geomorphological terms, as the task of extracting geomorphologically useful information from a three-dimensional shape. The information has been obtained either from field survey or from analysis of maps and aerial photographs. A distinction has been drawn between measurement of features that extend over large distances (macroform) and those that extend for only small distances (microrelief). Measurement of the latter is undertaken by field survey.

MACROFORM

In measurement of hillslope macroform most studies have focused on the attribute of gradient. Although this attribute is defined as the rate of change of elevation and is therefore a property of a point, it is normally measured as a property of an area (termed a *slope unit*) or a length (termed a *measured length*). The other property of hillslope macroform to have received some attention is plan form. This property is defined as the rate of change of aspect and hence, like gradient, is properly only defined for a point.

Field Survey

Field survey of hillslope macroform may attempt to provide areal information or be confined to measurements along transect lines. The most widely studied transect lines are those that follow the direction perpendicular to contours, generally termed *hillslope profiles*.

8

Areal Measurement of Gradient
The simplest way to obtain areal information about gradient is to select at random points over the hillslope surface and measure the gradient at those points, over some previously determined measured length. Although this method may seem a satisfactory way to characterise the gradient of hillslopes, Pitty (1969, p.9) argued that it is unrealistic to assume that the observations come from a single population, and that without making associated measurements at each sample point of aspect and proximity of the site to the summit or hillslope base, such sampling is of little use and unlikely to provide an adequate basis for numerical summaries of hillslopes.

Figure 2.1: A morphological map

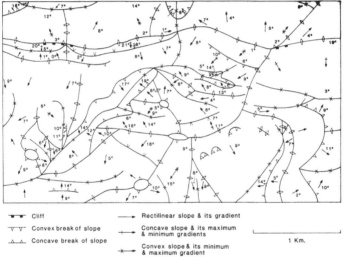

	Cliff		Rectilinear slope & its gradient
	Convex break of slope		Concave slope & its maximum & minimum gradients
	Concave break of slope		
			Convex slope & its minimum & maximum gradient

1 Km.

A more complete approach to areal measurement of gradient is provided by morphological mapping (Waters, 1958; Savigear, 1965). This method is contingent upon a major assumption, namely that hillslopes consist of planar and curved units that intersect at angular discontinuities, termed *breaks of slope*. In practical terms, it is also necessary that these discontinuities can be recognised by eye. Plane units (*facets*) have uniform gradients that are measured over their full lengths along the line of maximum slope.

9

Curved units (*segments*) do not have uniform gradient and for these units gradient is measured over short measured lengths at the minimum and maximum gradient for the unit. The product of this mapping technique is a morphological map showing the areal extent of each hillslope unit and the nature of the bounding discontinuities. For facets, the gradient and its azimuth are shown: for segments, the maximum and minimum gradients are shown together with the azimuth of the element (Figure 2.1). Although the broad principles of morphological mapping are well established, the practitioners of this technique have shown inconsistency in some of the details. Two procedures are worth special note (see Figure 2.2). Savigear (1965), in his description of the

Figure 2.2: Detail of morphological mapping showing discontinuities in the vertical plane and a lateral change along a hillslope from two facets to one concave segment (Key as for Figure 2.1)

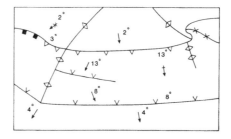

technique of morphological mapping, accepted that breaks of slope may not be perpendicular to the azimuth of maximum gradient of hillslope units. In extreme cases a break of slope can be parallel to the line of maximum gradient and thus mark a sharp change in aspect of the hillslope rather than a change of gradient. Savigear also allowed one break of slope to end without necessarily intersecting another. The most common example of this situation is where two facets pass laterally into a segment. Adherence to these procedures has not been universal in morphological maps produced subsequently by other workers (e.g. Gregory and Brown, 1966). Inasmuch as no attempt has ever been made to substantiate the assumption on which morphological mapping is based, these differences in application can be regarded as no more than

10

differences of interpretation. These differences would be resolved, and the usefulness of morphological mapping in hillslope studies would be greatly enhanced, if a statistical study were undertaken to test the assumption.

Hillslope Profiles
The commonest approach to field measurement of hillslope form has been surveying of hillslope profiles and, inasmuch as valley sides are the most widely-occurring type of hillslopes, valley-side profiles are sometimes regarded as synonymous with hillslope profiles (Cox, 1981). The approach suffers from limitations inherent in attempting to represent a surface by a line. In particular, it needs to be recognised that the line of maximum gradient may change its orientation along its length so that the vertical plane in which the profile line is defined (Young, 1974a) may have a sinuous plan form. Pitty (1966) considered that changes in orientation along the length of a profile line introduce a source of unknown variation into hillslope profiles and chose to discard all profiles that could not meet the condition of a near-constant bearing for the direction of maximum gradient. Although such an approach preserves the purity of the sample, it negates the possibility of representing all hillslopes by profiles and hence severely reduces the value of hillslope studies based upon the analysis of profiles. Other workers have considered that profiling can be used to characterise hillslopes irrespective of local variation in the azimuth of gradient or the curvature of contours. Young (1972) has presented a scheme for locating profiles to sample hillslope form (Figure 2.3). Although this scheme was devised for sampling within drainage basins, its basic principle of locating profile lines along a sampling line can be applied to all hillslopes.

A fundamental problem in measuring hillslope profiles is that of deciding where to terminate them. In morphological terms hillslope profiles end at gradient reversals. For valley sides these gradient reversals occur at the drainage divide at the upper end, and at the drainage line at the lower end. Since the drainage line, and very often the divide also, slopes downstream, adhering to the rule that the profile line must follow the line of maximum slope causes the profile to

Figure 2.3: Sampling scheme for hillslope profiles (*After Young, 1972,* Slopes. *Reproduced by permission of Longman, publishers)*

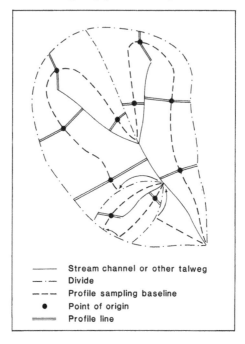

——— Stream channel or other talweg
— · — Divide
— — — Profile sampling baseline
● Point of origin
═══ Profile line

approach the endpoints asymptotically. Pitty (1966) proposed terminating valley-side profiles where the down-valley gradient equalled the valley-side gradient (see Figure 2.4). Such a solution is unsatisfactory if only because the hillslope profile thereby fails to record the morphology of the entire hillslope. Furthermore, in reality the line of maximum slope of most valley sides is neither perpendicular to the down-valley direction nor of constant orientation. In practice, it may be difficult to decide at which point a change in direction of a profile represents the beginning of the asymptotic curve rather than a legitimate change in direction. Young (1974a) proposed an extension of profiles to and across the divide (and drainage line) from the point at which the gradient along the profile line equals that of the axis of the divide (or drainage line) to that point on the other side of the divide (or drainage line) where gradient again equals that of the axis. The orientation of this extension should be perpendicular to the axis of the divide (or drainage line). Because the line of maximum slope of most valley sides is oriented slightly downvalley, this solution results in a dog-leg at each end of a profile line (see Figure 2.4).

Figure 2.4: Alternative solutions to the problem of terminating hillslope profiles. That proposed by Pitty (1966) is shown as (a); that proposed by Young (1974a) is shown as (b)

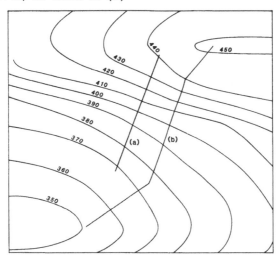

In areas where the upper ends of profiles are of low gradient (such as on plateaux) it may be difficult to identify the position of the gradient reversal or the drainage divide. In such cases Young (1974a) recommended extending the line of the profile until a consistent reverse slope is recorded. A similar difficulty may occur at the lower end, particularly where valley sides abut onto a wide floodplain. To continue the profile to the drainage line may present problems of two kinds. First, where there are levées or old drainage lines on the valley floor reverse slopes may be recorded before the main drainage line is reached. Thus, on morphological grounds, the profile must terminate before the drainage line. Second, in most surveys of hillslope profiles it is implicit that the forms being measured result from the operation of hillslope processes. If it

13

cannot be assumed that these processes are effective as far as the drainage line, then some criterion other than a morphological one needs to be used to decide where to terminate a profile. It is not at all clear what criterion can be used, other than the personal judgement of the surveyor, to identify the point at which hillslope processes cease to be the dominant control of land form.

The previous paragraph highlights the issue of the purpose of measuring hillslope form and the consequences that purpose has for decisions on measurement procedure. The purpose may be wholly for description. In this case the positions of the limits of hillslope profiles determine the area of land that is being described and the accuracy of the description. If profiles are terminated as Pitty proposed, then the description is of only part of the hillslope surface but it is a faithful description of that part. If Young's perpendicular extension method is adopted, the record is complete but some measurements give gradients that are gentler than any that actually exist on the hillslope. All measurements of profiles contribute to our overall knowledge of hillslope morphology and, hence, may contribute to our understanding of hillslope evolution. Bearing this in mind, it is important that measured profiles can be related to the operation of hillslope processes. They should, thus, follow the direction of sediment transport. Ending valley-side profiles at the divide and drainage line and aligning them along maximum slope accords with the nature of sediment transport by surface wash processes. However, subsurface water movement may be more affected by the pattern and direction of rock fissures such that the surface drainage divide is of little significance. Major mass movements, once initiated, may have sufficient momentum to transport sediment other than in the direction of maximum local surface gradient.

Because the technique of hillslope profiling has been so widely used it has been the focus of much of the methodological discussion concerning field survey of hillslopes, some of which is relevant in a wider context. It has already been noted that although gradient is a property of a point it is usually measured as the property of a length or surface. Most techniques of hillslope profile measurement are based upon measurement over a measured length of gradient or change in elevation. The commonest instrument used for surveys of hillslope profiles is the Abney level. The technique for using this instrument is described by Young (1974a). Other instruments for measuring gradient or change in elevation over measured lengths are described by Pitty (1968a), Blong (1972), Gilg (1973), Young (1974a), and Gardiner and Dackombe (1977). All these techniques require that the surveyor be able to stand on the line of profile. Churchill (1979) described a technique that can be used on precipitous gradients - an essential capability if field measurement of profiles is not to be restricted to a subset of the entire population of hillslopes. One method of measuring hillslope profiles that does not depend upon recording over measured lengths uses a profile recorder. This instrument, which was developed in the late 1960s, uses a damped pendulum to record on a rotating drum a continuous plot of the profile. Despite the potential of this instrument for rapid generation of hillslope profile data (Young, 1974a), no significant use appears to have been made of it.

One problem which has faced all surveys of hillslope profiles based upon measured lengths has been the choice of the size of measured lengths. In early work (e.g. Savigear, 1952) measured lengths were terminated at observed breaks of slope. Hence the profile was surveyed in units equivalent to transects across the hillslope units that form the building blocks for morphological maps. More recently (Young, 1974a; Parsons, 1978) it has been recognised that standardisation of measured lengths is necessary if comparability among data from different observers is to be achieved. Young (1974a) recommended a standard measured length of 5 m. Gerrard and Robinson (1971) investigated the effects of varying measured lengths by a factor of 4 on the same profile. Their results show clearly that as the size of measured length is reduced so the maximum recorded gradient increases but the effect of size of measured length on mean hillslope gradient is less clear and the recorded variation in mean gradient may be due as much to measurement error (which was not studied separately) as to any systematic changes associated with varying the size of measured length.

Measurement of Plan Form

Although hillslope plan form has been recognised
as a separate property of hillslope form and one
that may affect the operation of processes (Carson
and Kirkby, 1972, p.390; Young, 1972, p.177),
fewer measurements have been made of it. In
morphological mapping (Waters, 1958; Savigear,
1965) there is no explicit discussion of the plan
curvature of slope units comparable to that of
their profile curvature. For completeness, it
should be recognised that slope units may be
convex, concave, or planar in plan, just as they
may be in profile. Hence there are nine possible
shapes of slope unit (Figure 2.5). Fully to
describe hillslope form by morphological mapping
requires recording gradient and orientation of
slope units that are planar in profile and plan,
and the terminal gradients and orientations of
slope units that are curved in profile and plan.
Failure to recognise these nine possible shapes of
slope units is a weakness of the system of
morphological mapping described by Waters and
Savigear

Figure 2.5: The nine possible shapes for hillslope
units

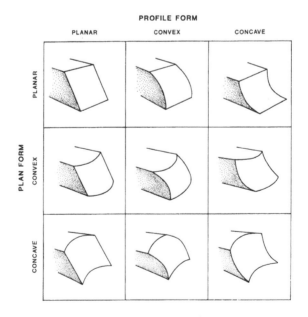

16

In surveying hillslope profiles, Pitty (1966) recognised the difficulties presented by variations in hillslope plan form and recommended restricting surveys to hillslopes that are plan planar. Young (1972, p.176) provided a procedure for surveying the plan form of hillslope profiles. In this procedure, the plan curvature of the profile line is measured by recording the angle subtended at a point on the profile line by lines from two ranging poles placed equidistant from the profile line and at the same height as the point on the profile line. The recommended distance to the two poles is 25 m, although this value has since been amended to 20 m (Young, 1974a p.37). This latter value was used by Parsons (1979). This technique of measuring plan form gives a value for a point which in some studies has been taken to be representative of a whole profile (Young, 1974a, p.37; Abrahams and Parsons, 1977) but elsewhere of a measured length (Parsons, 1979).

Measurements from Aerial Photographs and Topographic Maps

Although measurements of hillslope form can be obtained from aerial photographs and maps, the comparative loss of detail (as shown by Kertész and Szilárd, 1979) has led geomorphologists to be somewhat dismissive of their value (e.g. Young, 1972, p.147) compared to data obtained in the field. On aerial photographs and maps gradient is obtained by recording change in elevation over a measured length. Typically, these measured lengths are large compared to those commonly employed in field surveys.

Turner (1977) compared ten methods of gradient measurement from aerial photographs with the results obtained from a field survey using an Abney level and then evaluated all methods against a theodolite survey, which was taken as "absolute" control. Three methods of estimating gradient from the aerial photographs gave results comparable in accuracy to the Abney-level survey, although no one method achieved comparable accuracy over the full range of gradients examined. Superficially, these results are very impressive but since Turner does not specify the size of measured length it is difficult to evaluate the true viability of these methods of

17

obtaining data on hillslope form. Furthermore, the sites were selected so as to lie close to the principal points of the photographs and the photography was at a scale of 1:5000. It is likely that this study indicates the upper levels of accuracy of these methods rather than that which may be generally obtained.

Nowadays, most maps are prepared photogrammetrically so that measurements made from them can be no better than those from the source photography. A variety of techniques have been presented for obtaining average measures of hillslope gradient from maps (e.g. Rich, 1916; Wentworth, 1930) but the most useful for obtaining large numbers of separate measurements employ slope scales of the type described by Thrower and Cooke (1968). For the most part, data on hillslope gradient obtained from maps has been used to produce slope maps (e.g. Raisz and Henry, 1937; Calef and Newcomb, 1953; Le Roux, 1976) but more substantive geomorphic investigations have also been undertaken using such data (e.g. Strahler, 1956a; Ollier and Thomasson, 1957). The scope for measurement of hillslope form from maps is illustrated by the work of Chapman (1952) who used them to obtain a wide range of statistics on hillslope gradient (including mean, frequency distribution and orientation).

Recent developments in both cartography and remote sensing have enhanced the value of these sources of data on hillslope form. In cartography, digital terrain models provide ready sources of data. Elevation models may be used to provide automatic extraction of the same data as has previously been obtained manually from topographic maps (Engelen and Huybrechts, 1981). Alternatively, elevation models can provide (via Fourier transforms) models of gradient and ground curvature (Papo and Gelbman, 1984). Bell (1985) used a digital terrain model to draw profile lines on a topographic surface. By using this method, Bell was able to investigate the problems of sampling hillslopes using hillslope profiles much more easily than could be done from field survey. She identified substantial differences in the areal coverage of profile lines drawn from points spaced equidistant along Young's profile sampling line (Figure 2.6). The results of this study point to the weakness of profiling as a means of representing the three-dimensional form of hillslopes.

18

In remote sensing, altimetry using lidar and microwave altimeters is of potential use for measuring hillslope form. At the present time radar altimeters provide information on elevation

Figure 2.6: Areal coverage of hillslope profiles. Profiles were drawn from points spaced equidistant along Young's sampling baseline. Note the considerable non-uniformity in sampling density of interfluve and valley-floor areas (*After Bell, 1985, and reproduced by permission of the author*)

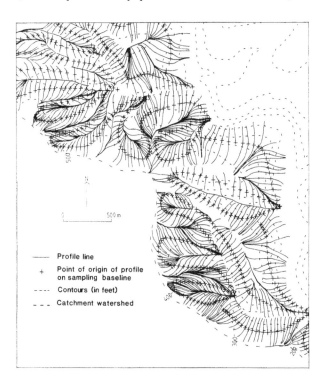

for relatively large areas. The pulse-limited footprint of the advanced altimeter to be included in the payload of ERS-1 is about 100m in diameter at a range of 10 km. It is fair to say that the value of such broad-scale data for hillslope studies is presently unknown as, indeed, is the precise nature of the relationship between ground surface morphology and the radar backscatter. Further research is needed into this topic.

There is little chance that in the foreseeable future developments in digital terrain models or remote sensing will be such as to provide sources of data comparable to that obtainable by field survey. This fact alone should not be sufficient to dismiss these methods as sources of useful information about hillslope form. The present paucity of field measurements of hillslope form indicates that this source may never yield sufficient information to allow a typology of hillslope forms to be established. Basic questions, such as whether form varies with climate, remain unanswered. Techniques that allow large amounts of data to be collected rapidly, even if the data do pertain to large measured lengths or slope facets, may have much to offer.

MICRORELIEF

Microrelief is the term used to describe small-scale elements of hillslope form. There is a widespread belief amongst geomorphologists that there exists a surface irregularity somehow superimposed upon the macroform of hillslopes and that the properties of this irregularity can be studied independently. This belief is expressed by Young (1972) who defined microrelief as "irregularities of the ground surface with dimensions of an order smaller than that of a slope sequence" (p. 201). Young considered that it is impossible to define microrelief in terms of dimensions, arguing that on the roughest hillslopes the microrelief may have dimensions larger than small valley sides (e.g. those of badlands). For the limited purpose of individual studies, a wide variety of dimensions have been used to characterise microrelief. Stone and Dugundji (1965) took it to comprise features having a vertical extent of between 8 cm and 3 m and a horizontal extent of between 12 cm and 19.2 m. Elsewhere, the term has been used to denote features with heights of as little as 2.5 cm (Dwornik, Johnson, Little & Walker, 1959). In contrast to the notion of identifiable microrelief, Savigear (1967) suggested that a continuous gradation exists in the size of features on hillslopes, an idea not dissimilar to the notion that hillslopes may be fractal surfaces (Scheidegger, 1961a). There has been considerable speculation regarding the value of fractals to geomorphology but, for hillslopes, recent work suggests that the fractal model may have limited applicability. Culling and Datko (1987) suggested that landscapes exhibit the properties of fractals over narrow, often overlapping bands. For the special case of scree slopes, Anderle (1987) argued that the fractal model is only superficially applicable. He showed that for these surfaces the relationship between fractal dimension and scale of measurement is a non-linear continuous function in the form of a series of overlapping bell curves. The peak of each bell curve was demonstrated to correspond to an individual scale of roughness that could be related to the size of particles forming the scree surface. Although the findings of this study cannot be readily extended to soil-covered hillslopes, they do point to the need for greater investigation into the scale relationships of hillslope form.

Whereas the majority of field surveys of hillslope macroform have been undertaken on hillslope profiles, this is not the case for measurements of microrelief. Studies have been more or less equally divided between attempts to obtain areal information on microrelief and those that have measured microrelief along transects. Campbell (1970) developed a frame to support 25 rods spaced at 25 cm intervals in a 1 m square to yield elevation data for 25 points within a 1 m^2 area. From these data Campbell constructed microtopographic contour maps. Other workers have obtained areal information by measuring profiles arranged in a grid (Crowther and Pitty, 1983), or radially (Stone and Dugundji, 1965). Where information has been measured along transects, microrelief meters generally similar to that described by Mosley (1975) have commonly been used (e.g. Simanton, Rawitz and Shirley, 1984; Abrahams, Parsons and Luk, 1988). In contrast to field measurements of macroform in which measurements of gradient have tended to be made directly (most commonly with an Abney level) and have usually been defined over measured lengths along the ground surface, measurements of microrelief have commonly been made in terms of differences in elevation and have typically recorded plane distance between observations. These distances have ranged from as little as 2.5 cm (Abrahams *et al.*, 1988) to as much as 1 m (Crowther and Pitty, 1983). Other instruments

21

appropriate for measuring microrelief include the automated profile meter described by Podmore and Huggins (1981) and Toy's linear erosion/elevation measuring instrument (Toy, 1983). The former has been designed for laboratory use but could be adapted for field measurements and has the advantage of producing digital output of elevation data. Toy's instrument is particularly suited to measuring short-term changes in microrelief.

To date there have been no recommendations regarding techniques for recording microrelief comparable to those given by Young (1974a) for measuring hillslope profiles. Microrelief remains an ill-defined and uncertain feature of hillslopes. There is urgent need for more quantitative work on this aspect of the measurement of hillslope form.

DESCRIPTION & CLASSIFICATION 3

DESCRIPTION

Although it is possible to describe hillslopes using purely qualitative terms such as 'convex' or 'conical', such an approach is of limited value. The terms are open to misinterpretation and do not allow accurate comparisons. A more restrictive approach is to define description as the task of converting measurements of hillslope form into quantitative statements. It is this approach that will be adopted here. Inasmuch as the majority of measurements on hillslopes have been of gradient so, too, does quantitative description depend heavily upon this attribute of hillslope form.

Gradient Frequency Distribution

One of the simplest ways to convert measurements of gradient into a quantitative description of hillslope form is to prepare a histogram of the frequency distribution of gradients recorded (Figure 3.1). This descriptive technique is as equally applicable to data collected from maps and aerial photographs as to that collected in the field. It may be used to summarise measurements of gradient made at locations scattered over hillslope surfaces (Strahler, 1956a) or those made along profile lines (Young, 1970) or those of hillslope units (Seret, 1963). A frequency distribution indicates the proportions of a hillslope falling within gradient classes, but it gives no indication of the shape of the hillslope. Indeed, radically different hillslope forms can give rise to similar frequency distributions of gradient (Pitty, 1968b). This weakness of

23

frequency distribution can, in part, be overcome
by sub-dividing the histogram to show, for
example, the frequency of gradients on upper,
middle and lower parts of hillslopes (e.g. Wilson,
1978).

Figure 3.1: Frequency distribution of hillslope
gradients

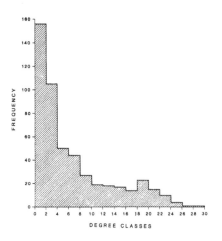

The characteristics of gradient frequency
distribution have been used to differentiate among
hillslopes (Tricart and Muslin, 1951) or to
identify peaks in the distribution, termed
characteristic angles (Young, 1961). Unfortunately, such interpretations of frequency
distributions have generally been undertaken
without the aid of statistical analysis and it is
questionable whether such features as peaks in the
distributions are representative of any more than
sampling error. In most studies the distribution
of gradient frequency has been found to be
right-skewed whether measured as an angle, tangent
or sine. This observation led Speight (1971) to
propose the tangent log-normal distribution as a
general model for hillslope gradient, and to claim
that where differences of distribution have been
found it has been due to inadequate sample size.
Speight argued that if all distributions of
gradient do tend to conform to this model then
differences in the shape of frequency distributions cannot be used to distinguish among
hillslopes and that the only statistics that do

24

have value for distinguishing among hillslopes are the mean and standard deviation of gradient. Furthermore, Speight claimed that departures from the log-normal model are rarely sufficient to justify recognition of characteristic angles. Notwithstanding Speight's claim other workers have attempted to interpret apparent characteristic angles. Molchanov (1967) provided a genetic interpretation arguing, for example, that alluvial surfaces have gradients less than 8° and that surfaces of contemporary colluvial downwash have characteristic gradients of 30°. Similarly, Carson (1975) has claimed that characteristic gradients may represent thresholds for mass movement processes.

Speight's own concept of a general log-normal distribution for hillslope gradients is challenged by the more recent findings of O'Neill and Mark (1987). These authors used digital terrain models to obtain maximum gradients of grid cells of 30 m side and acquired 160,000 measures of gradient from each of 18 locations. Their results show that no single transformation is capable of normalising all of the distributions. Within any area two major factors affect the shape of the gradient frequency distribution. First, the geomorphic character of the area strongly influences the development of areas of gentle slope, and second, the placement of the study area can affect the proportion of gentle slopes sampled. It is the proportion of land surface of gentle gradient included within a study area that determines the skewness of the gradient frequency distribution. Speight's result is due to studying areas all of which had large proportions of their surface inclined at gentle gradients.

Hillslope Profile Components

When using the technique of morphological mapping described in Chapter 2, it is necessary to assume that hillslope surfaces consist of slope units. Although this same assumption has sometimes been used as the basis for defining measured lengths in hillslope profile survey (Savigear, 1952), it can be avoided if a fixed measured length is used (Young, 1974a). Nevertheless, dating from the work of Savigear (1952, 1956), the assumption has underlain many attempts to describe hillslope profiles. Although the notion that hillslope

25

profiles could be regarded as being composed of a sequence of separate units was not new (Wood, 1942, had previously proposed a four-element model for hillslopes), Savigear separated the identification of slope units from any assumptions about their genesis, considering them to be no more than sections of the profile of either constant gradient (slope segments) or smooth curvature (slope elements). Young (1964) later narrowed the definition of slope elements to encompass only sections of the profile with constant curvature (ie. sections of the profile that are arcs of circles). In early work the identification of segments and elements from survey data was subjective which, given the problems of measurement error and hillslope irregularity, led to the possibility of differences of interpretation. Objective procedures for the identification of segments and elements were needed.

Ongley (1970) proposed a technique for the identification of segments based upon linear regression. From field survey data X and Y coordinates of the endpoints of measured lengths are calculated and these values form the data for the regression calculations. Starting at one end of the profile, a regression equation is fitted to the first three data points. If any of the residuals from the regression line is greater than a previously defined value (TOL), the first point in the sequence is dropped and a regression line fitted to points 2 to 4 and so on until none of the residuals exceeds TOL. Once three points are found that lie at .no greater distance from a regression line than TOL then this line is extended to the next and subsequent data points, again until such time as a residual greater than TOL is encountered. The procedure continues in this manner to test all data points in the profile. This technique yields output that is not always easy to use. Not all data points are necessarily assigned to segments, some may be assigned to more than one component and, as Cox (1983) pointed out, the results differ according to which end of the profile data is used as the starting point. Parsons (1973, 1976a) used a technique for the identification of segments that is also based upon linear regression. This technique assigns all data points to regression lines that extend for a minimum of three data points and yields output in which all data points are uniquely assigned to a segment. The most complete method for objective recognition of slope units is that developed by Young (1971) and termed *best units analysis*. In this technique Young retained the definitions of segments and elements as sections of the slope profile having uniform gradient and curvature, respectively, and merely provided the limits to variation in gradient and curvature that are acceptable within the concept of uniformity. Thus a section of a hillslope

Figure 3.2: Best segments, best elements and best units of a hillslope profile. For segments $V_{smax} = 10\%$; for elements $V_{cmax} = 25\%$

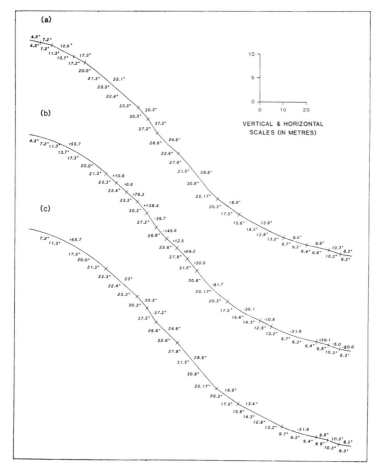

profile is considered to be a segment if the coefficient of variation of gradient (measured as an angle) does not exceed a specified limit. For general purposes Young recommended a value of 10% for this coefficient. Similarly, a section of a hillslope profile is deemed an element if the coefficient of variation of curvature is within a specified limit (25% for general purposes). Naturally, whatever values are chosen for the coefficients, some sections of profiles may satisfy both criteria. Where such alternatives exist, Young's procedure preferentially assigns measured lengths to the longer slope unit of which they form part. Where slope unit lengths are equal, measured lengths are assigned to the unit with the lower coefficient of variation and where both length and coefficient of variation are equal, measured lengths are preferentially allocated to segments. There are two variants of best units analysis, termed *best segments analysis* and *best elements analysis*, in which measured lengths are assigned wholly to segments and elements, respectively. The results of applying these techniques to hillslope profile measurements are shown in Figure 3.2.

All of these techniques for the objective recognition of slope units have weaknesses, as has been pointed out by Cox (1981), and those of Ongley's technique have already been referred to. Both Parsons's and Young's techniques do uniquely assign measured lengths to slope units, but both suffer from the weakness that the same units are not recognised when the analysis starts from the other end of the profile. Young's technique, like Ongley's, also suffers from the weakness that a spurious location of one survey station (in a small depression or on a vegetation mound, for example) may significantly affect the recognition of slope units. Weaknesses in these techniques indicate that as yet no satisfactory method for recognition of slope units has been found. Inasmuch as these slope units do form a convenient basis for quantitative description of hillslope profiles, further efforts in this direction are worthwhile.

An interesting method for describing hill-slope profiles based upon profile components is the nine-unit model presented by Dalrymple, Blong and Conacher (1968) shown in Figure 3.3. This model identifies nine possible components of a hillslope profile. With the exception of unit 1

28

which must always occur, any profile may consist of any number of the nine units in any order and any one of the nine units may be repeated any number of times. Clearly, the model is very flexible and permits description of any hillslope

Figure 3.3: Nine-unit hillslope model (*After Dalrymple et al.*, 1968, Zeit. für Geomorph.,12. *Reproduced by permission*)

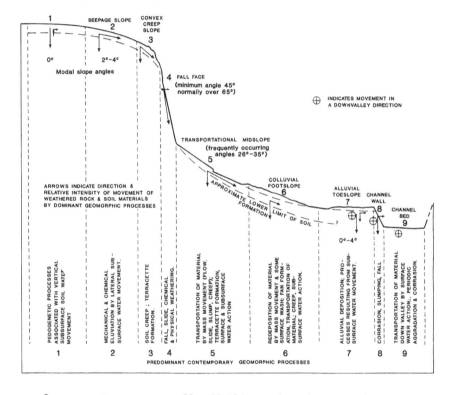

form. Its very flexibility is its weakness. Because all hillslope profiles can be regarded as realisations of the model it provides no means for distinguishing one hillslope from another. Its other weakness is to mix morphological description with genesis. Associated with each hillslope unit are generative processes. There is no certainty, for example, that all convex units on upper parts of hillslopes (unit 3) are caused by creep yet the model does not allow for any other generative process.

Fitting of Mathematical Functions

Any line or surface can be described by a mathematical function and the use of such functions to describe hillslope form can be dated back as far as the nineteenth century (Tylor, 1875). Such a description provides a convenient shorthand notation of hillslope form. The approach is more useful, however, if it is used either to examine how far hillslopes conform to some theoretical shape or to test alternative forms of description. Troeh (1965) argued that many hillslope forms could be fitted to the equation for a circular cone:

$$Z = P + SR + LR^2$$

where Z is the elevation of any pint on the surface, P is the elevation of the central point of the cone, R is the radius of the cone, S is the gradient at the apex of the cone, and L is half the rate of change in gradient along the radius of the cone. Troeh gave examples of hillslope forms that could be closely approximated by this equation, though offered little in the way of explanation for the fit. Tylor (1875) suggested that hillslope profiles approximate binomial curves on the grounds that running water will tend to produce a slope along which its velocity will be uniform. Lake (1928), however, measured nine hillslope profiles and found them to be better approximated by arcs of circles than parabolas. More recently, Bridge and Beckman (1977) investigated the possibility that concave profiles approximate the cycloid. The theoretical reasoning behind this investigation is that this curve is the least time path for a frictionless particle moving under gravity from one point to another not immediately beneath it. A second property of the cycloid is that several particles released simultaneously at different points on a cycloid slope will reach the lower extremity at the same time. Bridge and Beckman pointed out that such a profile would have the ideal shape for the most rapid dispersal of water, provided that friction was not important. Analysis of six profiles on basalt in the Darling Downs of southeastern Queensland showed the cycloid to be a good fit to upper parts of hillslope profiles characterised by shallow residual soils but that an exponential curve more closely approximated the lower parts of hillslope profiles where deeper,

30

transported soils were found.

The difficulty with comparisons of this type is that, whereas it is possible to identify a mathematical curve that has some particular property, it is seldom clear how far this property may be significant for the form of particular hillslopes. Two issues are confused. Bridge and Beckman recognised that it is not clear to what extent properties of curves that relate to the movement of frictionless particles also relate to those of real hillslopes on which the movement of particles is subject to frictional forces. Equally important is the fact that few, if any, hillslopes are formed wholly by one process. Thus the failure of hillslope shapes to conform to theoretical forms may be due as much to the complexity of processes that operate on real hillslopes as to factors not considered in the derivation of theoretical forms.

Fitting mathematical curves to hillslope profile data is one way of testing the notion that hillslopes are composed of slope units that intersect at discontinuities. If hillslope profiles are smoothly curved then low-order polynomials will fit the data as well as a sequence of slope units. Savigear (1962) fitted polynomial curves (up to fourth order) to measurements of cliff profiles and compared the goodness of fit to that of subjectively determined slope elements. The results were, however, inconclusive. In contrast, Young (1970), in a study of hillslopes in Brazil, found that a series of slope segments gave a consistently better fit than a third-order polynomial curve. Similarly, Abrahams and Parsons (1977) showed slope segments to be common among measurements of low-angled hillslopes in New South Wales.

On balance, the evidence suggests that hillslope profiles are not smoothly curved but that there are narrow zones within which changes in gradient or rates of change of gradient are concentrated. This conclusion is supported by Parsons (1977) who found hillslope segments to be more common than elements in data taken from a variety of locations. One reservation about this conclusion should be noted. All the studies which support it have obtained field data by recording gradient over measured lengths. Inasmuch as the basic data consists of, albeit short, rectilinear sections, there is the possibility that this biases analysis in favour of finding rectilinear segments. It would be useful to have the conclusion substantiated by data obtained from a continuous record of hillslope profile form such as that produced by the profile recorder.

Mathematical functions may also be fitted to parts of hillslope profiles. White (1966) argued that hillslope profiles measured in the area of Dayton, Ohio could be readily divided by eye into upper convexities, middle rectilinear elements and lower concavities. He fitted curves separately to each of these elements and examined the family of curves associated with each element.

Mathematical Transforms

An alternative mathematical procedure to describe hillslope form is the calculation of mathematical transforms, of which the Fourier transform is the most commonly applied. For a one-dimensional profile represented by points X_i with elevations Z_i, the Fourier transform may be given by

$$Z_i = \sum_{n=1}^{\infty} (\alpha_n \cos \frac{2n\pi X_i}{\lambda} + \beta_n \sin \frac{2n\pi X_i}{\lambda})$$

where α_n and β_n are the phase angles of the nth harmonics of the fundamental cosine and sine waves (of wavelength λ), respectively. Similarly, for a two-dimensional array of points X_i by Y_j with elevations Z_{ij}

$$Z_{ij} = \sum_{i=1}^{\infty} \sum_{j=1}^{\infty} \alpha_{nm} \cos \frac{2n\pi X_i}{\lambda_x} \cos \frac{2m\pi Y_j}{\lambda_y}$$

$$+ \sum_{i=1}^{\infty} \sum_{j=1}^{\infty} \beta_{nm} \cos \frac{2n\pi X_i}{\lambda_x} \sin \frac{2m\pi Y_j}{\lambda_y}$$

$$+ \sum_{i=1}^{\infty} \sum_{j=1}^{\infty} \gamma_{nm} \sin \frac{2n\pi X_i}{\lambda_x} \cos \frac{2m\pi Y_j}{\lambda_y}$$

$$+ \sum_{i=1}^{\infty} \sum_{j=1}^{\infty} \delta_{nm} \sin \frac{2n\pi X_i}{\lambda_x} \sin \frac{2m\pi Y_j}{\lambda_y}$$

Like the polynomial expressions, the Fourier transform may provide a convenient way of summarising morphological information. However, in this form it is of limited value for hillslope studies. There are two ways in which the Fourier transform is potentially of greater use. Analysis of the Fourier transforms from a variety of hillslopes may yield information on common frequencies of hillslope features, and whether a separate category of microrelief does exist. Of more immediate use, perhaps, is the fact that the power spectrum may be used to yield indices of hillslope form that can be employed in description (see the section following). To date there has been little exploitation of this technique for hillslope studies *per se*, rather applications have been in the general field of landform description (e.g. Horton, Hoffmann and Hempkins, 1962; Piexeto, Saltzman and Teweles, 1964; Stone and Dugundji, 1965; Rayner, 1972).

Indices of Hillslope Form

Indices derived from measurements provide a further means to describe hillslopes. Both Blong (1975) and Parsons (1978) have produced sets of indices that attempt to describe hillslope profile form. Both schemes subdivide profile form into four components: size, shape, gradient and roughness. The two schemes are compared in Table 3.1. It is interesting that both schemes independently identified the same four components of hillslope form, though neither author demonstrated any statistical justification for this breakdown. A principal components analysis of the data on Australian hillslopes used by Parsons demonstrates that the attributes assigned to the four components do show some tendency to load on separate components and that, using the conventional eigenvalue cutoff of 1, four principal components are recognised (Table 3.2). The first three components can be identified as steepness (component 1), shape (component 2) and size (component 3), but component 4 is more problematic. Although one of the measures of hillslope roughness loads on it (suggesting that the component may be identified as smoothness) the other does not. Further analysis of the correlation structure of hillslope attributes from a wider variety of environments is warranted.

Crozier (1973) developed morphometric indices to describe landslips and showed how each of the indices could be related to the processes that had caused the landslips. In this sense Crozier's study is not purely descriptive. Rather it attempts to show that morphometric data can be used to infer process mechanisms. Blong (1974) also devised indices of landslide morphology and

33

Table 3.1: Comparison of the indices of hillslope form used by Blong (1975) and Parsons (1978)

		(A) Blong	(B) Parsons
INDICES OF SIZE	(S1) (S2) (S3) (S4) (S5) (S6)	Total ground-surface length Vertical height Horizontal length Length of the straight line joining the two ends of the profile Ground-surface length from profile crest to point of maximum gradient S5 as a percentage of S1	(S1) Total ground-surface length (S2) Vertical height
INDICES OF SHAPE	(Sh1) (Sh2) (Sh3) (Sh4)	Crest curvature Basal curvature Profile mass Shape of the upper part of the profile	(Sh1) Profile curvature (Sh2) Position of maximum gradient as a percentage of S1
INDICES OF GRADIENT	(G1) (G2) (G3)	Mean gradient defined as arctan (S2/S3) Maximum gradient b coefficient of the linear estimating equation Y = a + bX fitted to profile survey data	(G1) Mean gradient defined as arcsin (S2/S1) (G2) Maximum gradient (G3) Percent profile length <2° - >-2° (G4) Percent profile length 2° - <5° (G5) Percent profile length 5° - <10° (G6) Percent profile length 10° - <18° (G7) Percent profile length 18° - <30° (G8) Percent profile length 30° - <45° (G9) Percent profile length >45° (G10) Percent profile length <-2°
INDICES OF SURFACE ROUGHNESS	(R1) (R2) (R3) (R4)	Number of changes of direction of curvature along the profile Value of S1/S4 Standard error of the estimate of the linear estimating equation Y = a + bX Value of S1/R1	(R1) Number of changes of direction of curvature along the profile (R2) Mean value of difference in gradient between adjacent profile segments

Sources: Blong (1975, pp.407-9) and Parsons (1978, p.435)

attempted to relate these to indices of hillslope form · (Table 3.3). His study identified few significant correlations between landslide variables and hillslope indices. The implication of this study is that hillslope form cannot be used to predict the occurrence and nature of mass movements.

The use of indices has proved particularly attractive in attempts to describe hillslope roughness and microrelief. Apart from their inclusion in the studies by Blong (1975) and Parsons (1978), there have been more specific studies on indices of microrelief and hillslope roughness (e.g. Stone and Dugundji, 1965; Schloss,

34

Table 3.2: Results of principal components analysis showing loadings after varimax rotation of the 16 indices used by Parsons (1978) to describe hillslope profiles in New South Wales

Index of profile form	Loading on all principal components that have eigenvalues >1			
	1	2	3	4
Total ground-surface length	-.181	.176	.890	-.061
Vertical height	.474	-.022	.837	.027
Profile curvature	.191	.803	-.060	.349
Position of maximum gradient	-.017	.797	.061	.005
Mean gradient	-.872	.312	-.170	.070
Maximum gradient	.879	.213	.111	.141
Percent $<2°$ - $>-2°$	.163	.420	-.018	.693
Percent $2°$ - $<5°$	-.245	.624	-.423	.290
Percent $5°$ - $<10°$	-.821	-.090	.308	.595
Percent $10°$ - $<18°$	.886	-.157	-.176	.077
Percent $18°$ - $<30°$	.204	.098	.308	.595
†Percent $30°$ - $<45°$	0.000	0.000	0.000	0.000
†Percent $≥45°$	0.000	0.000	0.000	0.000
†Percent $<-2°$	0.000	0.000	0.000	0.000
Changes of curvature	.844	.063	.079	-.109
Difference in gradient between adjacent segements	.239	.409	.238	-.682

† No part of any profile in this group had gradient $>30°$ or $<-2°$

Source: Parsons (unpublished)

1966; Hobson, 1972; Klein, 1981; Crowther and Pitty, 1983). Klein devised two measures of slope roughness of profiles; one is the ratio of the length of the profile line as measured to that of the straight-line distance between the two endpoints of the line and the other is the number of slope reversals along the profile line. Crowther and Pitty used the mean square of differences in angle between adjacent measured lengths along profiles. Schloss computed gradient and curvature statistics for points in a rectangular grid and used the changes that occurred in these statistics, when the cell size of the grid was increased, to describe roughness. Hobson also sought descriptive statistics of the roughness of surfaces defined by a grid of elevation values and proposed (i) bump frequency (i.e. the number of grid points that formed local

35

Table 3.3: Correlations between indices of hillslope form and landslide form

Hillslope indices	Landslide indices
Total ground-surface length (S1)†	Erosional slope length
Vertical height (S2)	Height of headwall
Distance of point of maximum gradient from hillslope crest as a percentage of S1 (S6)	Width of erosional zone
Crest curvature (Sh1)	Landslide circularity ratio
Hillslope mass (Sh3)	Gradient of shear plane
Mean gradient (G1)	Distance from hillslope crest to foot of shear plane
Gradient of erosional zone prior to failure	Percent shear plane covered by deposition
Surface irregularity (R4)	Volume of material deposited on shear plane as a percentage of total landslide volume

† Denotes the equivalent index in Table 3.1A
____ Denotes a positive correlation significant at p=0.001
---- Denotes a negative correlation significant at p=0.001

Source: Blong (1974)

high points on the surface - a two-dimensional equivalent of Klein's measure of slope reversals) and (ii) the distribution of the orientations of the triangular planes defined by three adjacent grid points. In Stone and Dugundji's study indices of the microrelief of surfaces were obtained from the high frequency components of a Fourier series transformation of hillslope data. Six indices were defined: (i) the average change in elevation, (ii) the average height of major microrelief features, (iii) the average steepness of microrelief features, (iv) the extent to which there is periodicity in the microrelief, (v) the irregularity of the surface, and (vi) the, so-called, cell-length, which delineates the distance from a given origin that must be traversed in order to encounter all significant microrelief features.

The widespread use of indices to describe microrelief probably reflects the fact that less attention has been given to understanding or explaining the microrelief features of hillslopes than to expressing their effect. In many studies of runoff and sediment yield from hillslopes it

36

has been recognised that these quantities are affected by microrelief, and quantitative relationships between them and indices of microrelief have been sought (e.g., Yair and Klein, 1973; Abrahams *et al.*, 1988).

Description of Hillslope Plan Form

It is not surprising that, given the paucity of measurements of hillslope plan form, there have been few attempts to describe it. Young (1972) proposed a five-fold classification based upon the value of the horizontal radius of curvature, R_h, as follows:

0 m < R_h <+50 m	Notably convex in plan
+50 m < R_h <+500 m	Slightly convex in plan
-50 m < R_h <0 m	Notably concave in plan
-500 m < R_h <-50 m	Slightly concave in plan
$\lvert R_h \rvert$ > 500 m	Straight in plan

Although this classification may be useful for describing the plan form at any point on a hillslope surface, the variation in hillslope plan curvature along profile lines as well as over hillslope surfaces restricts its utility. Parsons (1979) demonstrated that along hillslope profiles substantial variation in plan curvature is common and reported standard deviations of contour curvature P (measured in degrees/100 m) up to 64.6° in hillslope profiles up to 535 m long. Based on his findings that where the mean value of P ($\bar{P}$) >40 no part of the profile had P <0, and that where $\bar{P}$ -40 no part of the profile had P >0, Parsons proposed a simpler, three-fold classification of the plan form of hillslope profiles:

Wholly concave in plan	where $\bar{P}$ <-40
Wholly convex in plan	where $\bar{P}$ >+40
Convexo-concave in plan	where -40< $\bar{P}$ <+40

Troeh (1965) assumed that where hillslopes are curved in plan then the curvature can be described by an arc of a circle. The same assumption is implicit in the work of both Young and Parsons. This assumption has not been tested, although contour information on topographic maps could readily provide the data for such a test.

It was noted in the discussion of morphological mapping that Savigear (1965) considered discontinuities in the curvature of hillslopes to exist in plan as well as in profile. To date, there has been no demonstration of the validity of this assertion. Any attempt to test the assertion would require measurements of hillslope orientation along contour lines. Such surveys would be carried out in a fashion similar to hillslope profile survey. Along a line of constant elevation the azimuth of maximum slope would be recorded for sections (measured lengths) of the traverse. These data could then be analysed, for example using Young's *best units* program, to identify sections of the traverse along which plan curvature was constant or had a constant rate of change. Chapman (1952) presented analyses of hillslope orientation in a manner akin to polar diagrams used in sediment fabric analysis, recording both the gradient and orientation of hillslopes along traverses drawn on contour maps. Inasmuch as this type of analysis does not record the spatial arrangement of hillslope orientation measurements it fails to give a description of hillslope form. It is more useful as a summary description of topography.

Hillslopes as Fractal Surfaces

The discussion so far in this chapter has been concerned with techniques whereby sets of measurements made on hillslopes can be converted into descriptions. Savigear (1967) and Parsons (1978) have drawn attention to the distinction between the sets of measurements, which for hillslope profiles they termed *measured length profiles*, and the true form of hillslopes. This concern for the extent to which a particular set of measurements captures the true reality of hillslope form lies behind Gerrard and Robinson's (1971) consideration of choice of measured length in hillslope profile surveys. Implicit in these discussions is a belief that there is some optimum method of hillslope survey, and particularly choice of measured length, that will most closely capture the form of hillslopes. An alternative possibility is that hillslopes are fractal surfaces (Mandelbrot, 1975, 1977), as discussed in Chapter 2. If this were the case then the lengths of hillslope profiles measured on them would be a function of the measured length. Specific measures of hillslope profile length would depend upon the measured length and have no intrinsic meaning. Thus if the measured length were r then the number of measured lengths of that size necessary to survey the hillslope profile $N(r)$ would increase faster than $1/r$ so that the total length covered, $L(r) = rN(r)$, would increase without bound. Hence, $N(r) = r^{-D}$, where D is the 'Hausdorff dimension' of the profile (Mandelbrot, 1977, p.15ff). Mandelbrot suggested that for lines drawn on various geomorphic surfaces D $\simeq$ 1.2, a suggestion supported by the recent measurements of Culling and Datko (1987). The likelihood that hillslopes can be described as fractal surfaces remains an intriguing possibility and one that may have major implications for the manner in which geomorphologists attempt to describe hillslope form.

CLASSIFICATION

Classification - the grouping of entities into classes - is fundamental to scientific enquiry. Even so, there has been remarkably little effort expended in classifying hillslope forms, *sensu stricto*. Parsons (1973, pp. 5-6) argued that the emergence of geomorphology in the latter part of the nineteenth century led to an emphasis within the discipline on landform genesis. In consequence classifications of hillslopes formed part of genetic interpretations of landscape (e.g. Davis, 1899) rather than being classifications of morphology. Inasmuch as classification is a prelude to understanding (Jevons, 1887), conflation of the process of classification with that of genetic interpretation must result in confusion and may well explain the comparative lack of progress towards understanding hillslope form (Parsons, 1973 pp. 8-9). In an extreme way, Penck's analysis of hillslope form (1924) reverses the sequence of classification and understanding by deducing the forms that result from presumed processes and then using the existence of these forms in the landscape as a means of identifying the processes.

Of the studies that have attempted to account for the hillslope forms observed in particular localities, few have separated the process of classification from that of genetic interpretation. Those that have seem to have regarded classification as a relatively trivial task. Seret (1963) undertook an analysis of the gradient frequency distribution of a limestone shale area in Belgium and identified peaks in the distribution. He used these peaks to determine gradient class boundaries and then mapped the distribution of hillslope components in these classes. Subsequently, he provided a genetic interpretation of each of the gradient classes. Seret's study does not attempt to classify or explain hillslopes as such but rather only their components. The organisation of these components into hillslopes does not form part of the study. Young (1970) also used hillslope components as the basis of his classification of hillslope profile forms in the Xavantina-Cachimbo area, Brazil. He first used best units analysis (Young, 1971) to determine the components and then identified six classes of hillslope profile form on the basis of component composition and association. A somewhat different approach to classification was taken by Woods (1974). This author fitted polynomial curves to hillslope profiles and then used the degree of the polynomial as the basis of classification, identifying three classes (first, second and third degree) in his study area. Both Young and Woods, in the latter parts of their studies, used their hillslope profile classes together with the spatial distribution of members of the classes as a means to infer genesis and evolution.

None of the authors of these studies indicated difficulty in achieving classification. Woods, for example, simply used an F-test to determine which degree of polynomial should be used to fit his hillslope profiles. Such a procedure will lead to clear-cut classes even if the hillslope profiles lie along a continuum of degree of curvature, a possibility that Woods did not discuss. Young recognised that not all members of a class of hillslopes would be identical and commented upon the range of forms within classes. However, he provided no formal testing of between- and within-class variance that might have been used to substantiate the class divisions. In contrast, studies that have been concerned wholly with the problem of classification of hillslope form have been rather less unequivocal in their conclusions.

Component-based Classifications

Attempts to classify hillslopes on the basis of their constituent components can be divided into two groups; deterministic and stochastic.

Deterministic classifications

Ahnert (1970b) used the number of constituent components to classify hillslope profiles and identified *simple profiles* (those of which more than 90% of the length is occupied by one component), *compound profiles* (those containing two or three components) and *complex profiles* (those containing more than three components). Since each hillslope component may be rectilinear, convex or concave, there are three types of simple profile, six types of two-component compound profiles and twelve types of three-component compound profiles, and so on. Savigear (1967) developed a similar classification of hillslope profiles and defined *primary hillslope profiles* as being those whose components are all of one type (ie. all segments, all convex elements, or all concave elements), *secondary hillslope profiles* as those containing two types of component and *tertiary profiles* as those comprising all three types of component. Within any hillslope profile Savigear defined as *consociations* groups of components that are generally convex or generally concave, and as *associations* groups of components

Figure 3.4: Classification of hillslope component groups (*After Savigear, 1967,* L'Evolution des Versants, *P. Macar, ed. Reproduced by permission*)

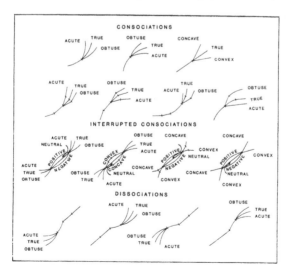

41

that are generally convexo-concave or concavo-convex. As with Ahnert's scheme, the number of possible hillslope profiles or component groups is large (see Figure 3.4). Savigear showed seven types of consociation and four types of association. A more formal presentation of similar ideas has been made by Demirmen (1975) in a more general discussion of profile analysis. This discussion was linked to a computer-based procedure for classifying profiles on the basis of their composition of straight and curved sections.

Because of their complexity, it is unlikely that schemes of this type can be useful for classification of hillslope profiles. In terms of ordering the complexity of hillslope forms so that they can be better comprehended (and ultimately explained) schemes such as these do not seem to have much value.

Stochastic classifications
Stochastic classifications attempt to avoid the problem of an excessive number of classes by regarding a sequence of hillslope components as a realisation of some stochastic process. Whereas differences among some sequences may be no more than the random variation associated with the expression of the stochastic process, other differences may be due to differences in the generating stochastic process. Parsons (1976a) examined the possibility that the sequences of components that comprise hillslope profiles may be realisations of Markov processes (see Figure 3.5) and sought to identify differences in the underlying processes as a way of distinguishing groups of hillslopes. The study failed to identify differences of this type suggesting that all the hillslopes studied could be modelled as realisations of the same process. Such a conclusion undermines analytic classifications of the type proposed by Ahnert, Savigear and Demirmen. The Markov model is the only stochastic model so far applied to the description of sequences of hillslope components and for the reasons discussed by Thornes (1973) may not always be an appropriate one. There is a need for analysis of sequences of hillslope units using other, and perhaps more physically-based, stochastic models. The full potential of this approach to hillslope description remains to be discovered.

42

Figure 3.5: Markov model for hillslope profiles (*After Parsons, 1973*)

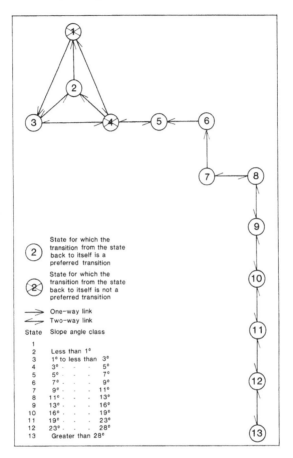

Classification by Indices of Morphology
Indices of hillslope form of the type developed by Blong (1975) and Parsons (1978) can form the basis of classification. Both of these authors used their indices in numerical classifications of hillslope form (Figures 3.6 and 3.7). This technique allows a more rigorous examination of between- and within-group variance so that classes can be justified on statistical grounds. In applying this approach to the hillslopes studied by Young (1970), Parsons was able to show that only four classes could be substantiated on

statistical grounds. Furthermore, he was able to point to different degrees of homogeneity of the classes.

Figure 3.6: Dendrogram for hillslope profiles. Profiles are grouped according to their level of similarity as measured by Parsons's indices of hillslope form given in Table 3.1. A statistical test supports identification of 3 classes and 7 unclassified profiles at a similarity level of 0.75 (*After Parsons, 1978*, Inst. Br. Geogr. Trans., New Series, 3. *Reproduced by permission*)

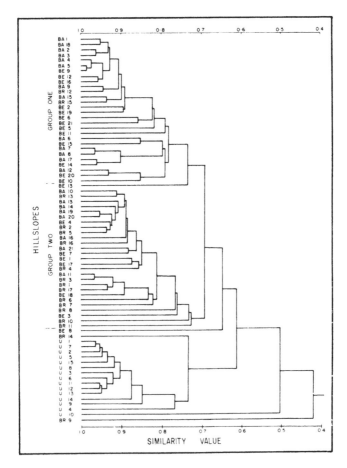

Numerical procedures of classification focus on significant issues of hillslope classification,

Figure 3.7: Hillslope types and classes. Profiles of the members of two of the groups from Figure 3.6 are shown and indicate both between-class differences and within-class variability (*After Parsons, 1978*, Inst. Br. Geogr. Trans. New Series, 3. *Reproduced by permission*)

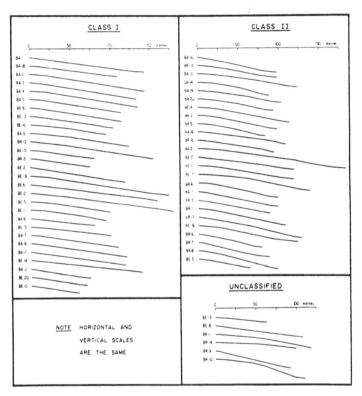

namely those of what entities are to be classified and how to measure their similarity. To date, these classifications have been of individual hillslope profiles. To the extent that these profiles may be regarded as no more than convenient, if unsatisfactory, ways of representing the form of three-dimensional hillslopes, classification of them on an individual basis may not be particularly useful. The results may have more to say about the problems of using a two-dimensional profile to sample a three-dimensional form than providing any insight into understanding hillslope form. The issue of sampling and sampling error needs to be addressed separately from the question of classification.

Both Blong and Parsons defined a set of attributes of hillslope form which they could use as the basis of classification. The two sets are similar but not identical. Doubtless other attributes could be identified to add to a combined list. Mosley and O'Loughlin (1980) criticised the choice of Parsons's attributes, particularly pointing to the large number of attributes measuring the property of hillslope gradient. The problem of choice of attributes on which to measure morphological similarity is not unique to the study of hillslopes: indeed it is well documented in the field of biology (see Sneath and Sokal, 1973, chapter 3). The general recommendation given there (p.106), of the need for a large set of attributes, applies equally well to the classification of hillslope form.

HILLSLOPE FORM   4
& CLIMATE

Several factors interact to determine the form of an individual hillslope, but it is probably climate that has been the subject of greatest investigation. To some extent, of course, this interest in the role of climate stems from, and is part of, climatic geomorphology. The notion that different landform assemblages can be associated with different climates can be traced back to Davis's writings on the temperate fluvial cycle (1899), the arid cycle (1905) and the glacial cycle (1900, 1906). In the more recent past German and French geomorphologists (e.g. Büdel, 1948, 1963; Tricart and Cailleux (1965) have been particularly active in emphasising the role of climate in landform development. Within this general context some, though perhaps remarkably little, attention has been given to the question of climatic influences on hillslope form. For the broad category of sub-aerially formed hillslopes "surprisingly little objective morphometric evidence exists .... and the recognition of distinct landform assemblages has depended less on total landscape morphometry than on the occurrence of less frequent but more spectacular type-landforms, such as inselbergs or pediments," (Stoddart, 1969, p.174)

The issue is a complex one involving several questions whose separate identity has not always been recognised. These questions are as follows:

(a) Does the form of hillslopes differ from one climatic zone to another? If so, is this because:
(i) climate affects hillslope form directly (i.e. by modifying the effects of processes) so that if all other factors

are held constant, differences in hillslope form will be observed solely as a result of moving from one climatic zone to another, or
(ii) climate affects other factors controlling hillslope form (e.g. surface materials, drainage density) so that the effects of climate are manifested only in an indirect way?
(b) Do the processes that operate on hillslopes themselves differ from one climatic zone to another? If this is the case, then do these differences in processes result in different hillslope forms?
(c) Does the manner of hillslope development differ from one climatic zone to another?

In these questions two terms require closer definition: hillslope form and climatic zone. If the relationship between hillslope form and climate is being investigated then it must be made clear what is meant by hillslope form. The term might refer to some notion of 'average' hillslope form within a specified climatic zone or be taken to imply a suite of hillslope forms characteristic of the zone. Equally, interest may focus on what may be termed differences of macroform (i.e. differences that are manifest at the scale of the whole hillslope), or it may focus on component parts of a hillslope (for example, differences in the concavity of footslopes), or it may deal with differences of microrelief, surface irregularity, and the like. The notion of a climatic zone is also scale-dependent. It is probably appropriate to recognise two scales of climatic zone. One is the local scale, concerned with differences in microclimate, particularly aspect. At this scale the relationship of hillslope form to climate has been examined within the context of valley asymmetry. The other scale is the regional one. At this scale relationships have been sought between hillslope form and regional climate.

These two scales of climatic variation provide the major subdivision of this chapter. In the first part the questions of the variation in hillslope form with respect to regional climate will be examined. The second part will consider valley asymmetry.

VARIATIONS IN HILLSLOPE FORM WITH REGIONAL CLIMATE

Toy (1977) considered the question of whether differences in hillslope form could be identified solely as a result of differences in climate. He designed an experiment to evaluate differences in hillslope profiles measured at different locations along two transects within the United States. One transect followed a line of latitude, hence differences along this line could be related to variation in total annual precipitation. The other transect followed a line of longitude, and hence differences in mean annual temperature. He attempted to hold other variables constant by selecting, along these transects, sites where hillslopes had developed on approximately horizontally bedded marine shales and were south-facing. Regrettably, as pointed out by Dunkerley (1978), the design of the experiment was flawed. Toy failed to identify all of the variables that may separately contribute to control hillslope profile form (Dunkerley identified in addition hillslope plan form and evolutionary stage). Thus all influences other than climate were not either held constant or statistically controlled for. Furthermore, Dunkerley noted the discrepancy between the length of the climatic record (1951-1971) used in the analysis and the likely period of time over which the hillslopes had developed. Given these shortcomings (and the fact that Toy misinterpreted the results of some of his statistical analysis) it is difficult to reach any conclusion about Toy's study other than that it indicates the near impossibility of establishing directly a relationship between hillslope form and regional climate. With the possible exceptions of badlands and anthropogenic hillslopes (where the absence of vegetation may be a controlling factor in any case), it seems unlikely that a sample of hillslopes taken from a wide range of climatic regions, but which have developed within a short period of known climatic stability, can ever be found. Hence it may be impossible to satisfy the conditions necessary for the type of direct test of the control of regional climate on hillslope form envisaged by Toy.

The contrary view, namely that climate does not directly affect hillslope form, has been most clearly expressed by King (1953, 1957, 1962) who has claimed that "to introduce climate as a controlling force in landscape-making is ... to include inference with fact" (King, 1962, p.156) and that "similar hillforms may be found under like conditions of bedrock and relief in all climatic environments short of glaciation or wind-controlled, sandy deserts" (King, 1957 p.81). The similar hillform to which King specifically alluded is the four-element hillslope (Figure 4.1) first identified by Wood (1942). A similar stand has been taken by Ruhe (1975). Frye (1959) took King (1957) to task and pointed out that the fact that the same four elements could be recognised in hillslopes developed on the same lithology but located within different climatic regions is not the same as arguing that hillslope form is independent of climate. Given that the relative size of each element may change, considerable variation in hillslope form may be encompassed without invalidating the model. Hence it is scarcely surprising that the model fits hillslopes in a wide range of climates. The four-element hillslope is probably best seen as providing a framework within which to describe hillslope forms in different climatic settings (e.g. Carson and Kirkby, 1972).

Figure 4.1: Four-element hillslope (*After King, 1957,* Trans. Edinburgh Geol Soc.,17. *Reproduced by permission)*

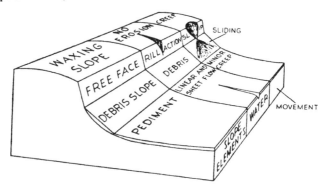

Lithology introduces a further complication into evaluation of the effects of climate on hillslope form (as illustrated by Wirthmann, 1977). It is worth noting that Toy undertook his study on hillslopes developed in marine shales whereas the best examples of four-element hillslopes are found in resistant strata. It cannot be denied that some lithologies exert a strong influence: bornhardts, for example, are characteristic of granites in a wide range of climatic environments. The results of studies of the variation of hillslope form with regional climate may well depend in large part on the lithologies on which the hillslopes under study have developed.

Hillslope Basal Concavities and Climate

Fair (1948b) drew attention to the differences between hillslopes of the semi-arid Karroo and those of the sub-humid interior of Natal (Figure 4.2) in the sharpness of the concavity of what he termed the "transitional" slope (i.e that between

Figure 4.2: Contrasts in the sharpness of piedmont junctions between semi-arid Karroo and sub-humid Natal (*After Fair, 1948b*, S. African Geog. J.,30 *Reproduced by permission*)

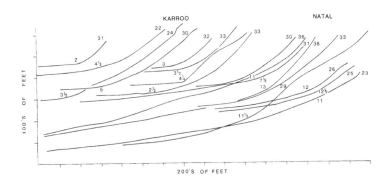

a talus slope and a pediment). Oberlander (1974) identified a similar difference between the hillslopes of the Mojave Desert and those of the Sonoran Desert. Abrahams, Parsons and Hirsch (1985) have argued that if it can be assumed that the principal agencies fashioning these concavities are hydraulic processes then a suitable expression to relate the shape of the hillslope to hydraulic processes is

$$S \propto L^{l/n}D^{p \cdot n}/X^{-m \cdot n}$$

where S is the gradient at a point on the concavity, L is the sediment load being

51

transported past that point, X is its distance from the divide, D is the mean size of particles mantling the surface at the point, and the exponents m, n, and p are all positive. In arid and semi-arid areas where there is a strong relationship between gradient of hillslopes and the size of particles mantling their surfaces (Abrahams *et al.*, 1985) D is an important variable in this equation. Specifically, as D decreases downslope so does gradient and the rate of change of gradient across hillslope basal concavities is relatively great. In more humid areas, by contrast, the surface material contains a greater proportion of fines and the relationship between particle size and gradient is weak or absent. Hence D changes little along the length of hillslopes. As a result S decreases more slowly across basal concavities.

Duricrusts

Duricrusts are hard, massive layers that are thought to form within the regolith by pedogenic processes active in tropical environments. Typically duricrusts form in horizontal or near horizontal sheets of very variable thickness. It has become common practice to label duricrusts according to their cementing agents: hence silcrete (quartz), calcrete (calcite), gypcrete (gypsum) etc. The importance of duricrusts for hillslopes lies in their resistance to erosion. Although they are believed to form in low-lying plains, subsequent erosion leaves the duricrusted areas upstanding, protected from erosion by the duricrust which acts as a cap rock (Figure 4.3). The typical hillslopes that surround the duricrusted surface have been termed *breakaways*. They consist of an upper cliffed section, whose height is closely related to the thickness of the duricrust, underlain by a concave section formed within the softer underlying material. This concave section is often mantled, particularly in more arid climates, by fragments derived from the duricrust (Figure 4.4). Hillslopes whose profile form is controlled by a duricrust are indicative of a past climatic setting, although they are now found in a wide range of locations, e.g. Nigeria (Moss, 1965; Dowling, 1968), Uganda (Pallister, 1956a), Brazil (Young, 1970), Cyprus (Everard, 1963), Egypt (Dumanowski, 1960), South Australia

52

(Ollier and Tuddenham, 1962). It is worth noting that hillslopes that owe their form to duricrusts closely match the four-element model of Wood

Figure 4.3: Hillslope with a duricrust caprock

(1942). Their present widespread occurrence may have contributed to King's (1957) contention that hillslope form is unrelated to climate and indicates the weakness of examining the relation of hillslope form to the present climate.

Figure 4.4: Profile of duricrust-capped hillslope

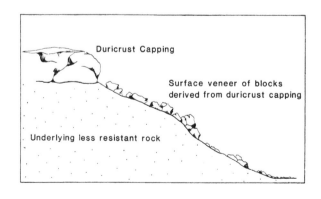

53

Hillslope Processes and Climate

It is only recently that the number of studies of hillslope processes has become sufficiently large for any analysis of their variation with climate to be attempted. Young (1972, 1974b) presented what evidence there was to indicate the effects of climate on the rates of some hillslope processes, but it is only with the more comprehensive data source of his later publication (Saunders and Young, 1983) that it becomes possible to attempt to identify climatic controls on the effectiveness of hillslope processes. Figure 4.5 shows the

Figure 4.5: Variation with climate of rates of soil creep (measured in mm/yr) and surface wash (measured in Bubnoff units B). Soil creep rates shown by pecked line; surface wash rates shown by continuous line (*Based on Saunders & Young, 1983*, Earth Surface Processes and Landforms, 8, Copyright John Wiley & Sons Ltd. 1983)

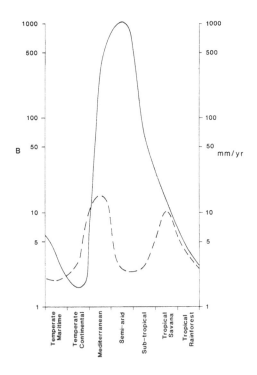

variation with climate in the averages of measurements of rates of soil creep and surface wash on hillslopes gentler than 25°. The graphs show not only considerable variation in the rates of individual processes with climatic region but, notwithstanding the difficulty of comparing linear and volumetric measures, indicate changes in the relative importance of these two processes.

Differences in the *absolute* rates at which hillslope processes operate show how fast hillslopes develop in particular climates and, given the phenomenon of climatic change, indicate the relative importance of particular climates for hillslopes that have developed under a variety of climates. It is differences in the *relative* rates that may be significant for the relationship of hillslope form to climate. If a particular hillslope form can be associated with a particular set of process rates then these results, indicating that relative rates of processes do differ with climate, would imply that hillslope form varies with climate. However, it has to be recognised that data on which this conclusion is based are not unbiased. No individual study of hillslope processes has sought to obtain representative rates of particular processes for a climatic region. Rather it is likely that the sites of measurements will have been biased towards those localities at which the process is most active. However, given the ubiquity of soil creep and surface wash and the fact that steep gradients and sites on unconsolidated rock have been excluded, it is probable that the values given in Figure 4.5 are close to the averages for their climatic regions. The same is unlikely to be true for the data on solution loss and landsliding also presented by Saunders and Young, and these results have not been considered here.

The Effect of Climate on Hillslope Evolution

In the debate that followed the emergence of the rival Davisian and Penckian models of hillslope evolution it was sometimes claimed that slope decline is more characteristic of humid environments whereas parallel retreat (as Penck's model was incorrectly termed) dominates in more arid climates. Little of substance emerged from this debate and only more recently has the issue been usefully addressed within an attempt to examine the effect of climate on the balance of hillslope processes, and the consequences of changes in this balance for hillslope evolution (Kirkby, 1976a). In his study, Kirkby used a process-response model to examine the effect on the development of an initially straight, 26.5° hillslope profile of varying the total annual rainfall, mean rainfall per rainday, and mean daily potential evapotranspiration. These data were used to determine the relative importance of overland and subsurface flow and hence the effects of surface wash and chemical removal. The results of this modelling for different annual rainfalls are shown in Figure 4.6. This theoretical study indicates that, if all other variables are held constant, as rainfall increases hillslope profiles become more convex and that the location of downcutting becomes, accordingly, more concentrated near the divides.

Figure 4.6: Comparison of simulated hillslope profiles after 100,000 years, for different annual rainfalls. For all profiles rainfall/rainday = 15 mm and mean annual potential evapotranspiration = 2000 mm (*After Kirkby, 1976*, Geomorphology and Climate, *E. Derbyshire ed.. Reproduced by permission of John Wiley & Sons Ltd.* Copyright 1976)

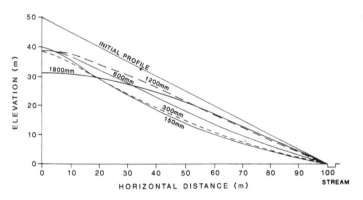

Hillslope Microrelief and Climate

Hillslope microrelief varies with climate most obviously as a result of the variations in soil thickness that accompany climatic differences. In humid climates, where soils tend to be both continuous and deep, microtopographic variations are less pronounced, whereas in more arid climates, where the surfaces of hillslopes are more commonly formed of bedrock or strewn with rock fragments, microrelief is more evident. No study has been based on this everyday observation so that no quantitative relationships between hillslope microrelief and climate have been identified. However, several microrelief forms have been associated with climate-specific process environments. Those of periglacial, semi-arid and humid temperate areas are examined in this section.

Periglacial Environments
Hillslopes cut in resistant bedrock commonly exhibit a stepped profile (Figure 4.7), particularly where the bedrock is bedded or well jointed. Small rocky cliffs alternate with

Figure 4.7: Hillslope profile in a periglacial environment, showing frost-riven cliffs (1), tors (2) and altiplanation terraces (3) (*After, Czudek, 1964*, Biul. Peryglac.,14 *Reproduced by permission*)

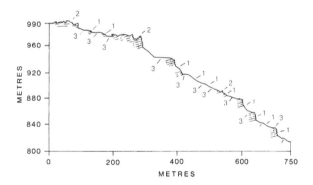

debris-covered benches, termed *altiplanation terraces*. On the cliffs frost shattering preserves the rocky appearance whereas on the benches freeze-thaw beneath winter snow patches is believed to maintain the angle between the step and the riser (Waters, 1962; Czudek, 1964). Weathering beneath patches of snow (termed *nivation*) is also thought to be responsible for the more irregular hollows (*nivation hollows*) that are commonly found on hillslopes in periglacial environments (Lewis, 1939). On hillslopes covered

57

in a mantle of soil or rock debris, processes of
solifluction and sorting may lead to distinctive
microrelief forms. Sharp, (1942) described stone-
banked lobes and steps on hillslopes of 10° to
25° formed in material that has moved downslope by
solifluction (Figure 4.8). The treads of the

Figure 4.8: Hillslope microrelief in a periglacial
environment due to solifluction and particle sorting
processes (*After Sharp, 1942,* J. Geomorph.,5.
Reproduced by permission of the author)

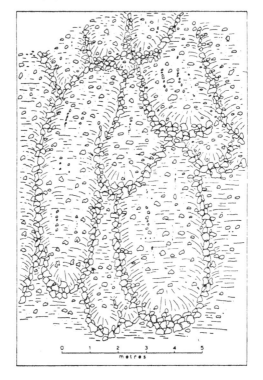

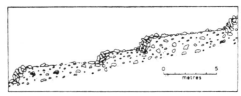

steps and the central parts of the lobes are underlain by fine-grained material whereas the risers and frontal parts of the lobes are stony. The formation of the stony banks is poorly understood but sorting of the coarser fragments by frost heaving was thought by Sharp to be a factor. Examples of these features have been found with risers as high as 5 m and steps as wide as 30 m (Galloway, 1961).

Semi-Arid Environments
Sorting in the soil layer to yield a stepped hillslope microrelief is also characteristic of semi-arid areas. Soils rich in expanding clays such as montmorillonite undergo swelling and heaving when wetted, and shrinkage and cracking on drying. Sorting of the stone layer leads to concentration of the coarse fragments in a riser on the downslope side of the expanding clay mass which tends to form into a relatively stone-free tread (Figure 4.9). Ollier and Tuddenham (1962) have termed this microrelief *desert gilgai* but Mabbutt (1977) suggested the term *sorted stone steps* to emphasise the analogy with periglacial microtopographic forms.

Figure 4.9: Hillslope microrelief in a semi-arid environment due to wetting and drying of expanding clays and particle sorting processes (*After Mabbutt, 1973*, Lands of Fowlers Gap Station, New South Wales, *J.A. Mabbutt ed. Reproduced by permission of the author*)

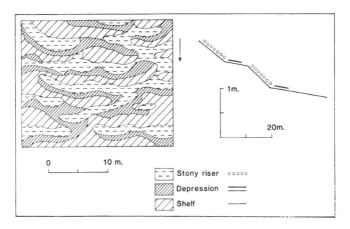

59

Climate

Humid Temperate Environments
The most widespread microrelief forms on moderate
to steep hillslopes of humid temperate areas are
terracettes (Figure 4.10). These features have
usually been identified on grassland hillslopes
though this may reflect the ease with which they
can be seen under this type of vegetation rather
than indicating any vegetational control. Despite
their common occurrence, there is no general
agreement as to their origin. In a review paper,

Figure 4.10: Hillslope terracettes in a humid
temperate environment

Vincent and Clarke (1976) showed that although
many authors believed terracettes to be
contemporaneous with the humid temperate climates
in which they are now found, others attributed
them to processes acting under former periglacial
climates. The role of animals is not clear. The
regularity of terracettes is difficult to explain

60

by any hypothesis that favours formation by animals. Part of the debate on the origin of terracettes may be due to an incorrect assumption that all terracettes are alike. Anderson (1972) distinguished what he termed *normal terracettes* and *tear terracettes*. The former have gentler and wider risers and gentler and longer treads than the latter. A second important difference is that the risers of tear terracettes are generally lacking in vegetation. Higgins (1982) identified a third type of terracette which he termed a *grazing step*. This terracette consists of turfed risers but almost flat, narrow treads that may be bare of vegetation. Like Anderson's normal terracette, this type has considerable continuity across the hillslope. Higgins showed that grazing steps can form rapidly (in a few months) and, as their name implies, are due to animal disturbance. Chemekova and Chemekov (1975), likewise, distinguished animal tracks with bare treads from what they termed *delapsive terracettes* which have bare risers but vegetated treads.

LOCAL CLIMATIC CONTROL OF HILLSLOPE FORM

Although valley asymmetry is not synonymous with the effects of local climatic variation on hillslope form, the majority of studies of asymmetrical valleys have been concerned with climatic explanations for observed differences. Valley asymmetry thus forms a convenient framework within which to examine the effects of small-scale climatic differences on hillslope form and, equally, the context of climatic control of hillslope form provides an appropriate opportunity to discuss valley asymmetry.

Aspect-Related Differences in Valley-Side Gradient

The majority of studies of valley asymmetry have examined differences in gradient between opposing valley sides. At first sight it might appear that the conclusions of these studies are both diverse and contradictory. Whilst many authors have found north-facing hillslopes to be steeper, there are studies which have shown the steeper hillslopes to face south, east or west. These varied results may appear more understandable if the orientation of the steeper valley side is plotted against the

61

Climate

latitude of its location (Figure 4.11). Notwithstanding the reservations expressed by Kennedy (1976) about recognising a typology of valley asymmetry, it would seem that some latitudinal grouping of the orientation of the steeper valley side does exist. In the mid-latitude zone (approximately 30°-45°N) all studies of asymmetry have found northerly facing valley sides to be the steeper ones. In the more poleward latitudes both north- and south-facing steeper valley sides have been identified. South of latitude 30°N east- and west-facing hillslopes have been found to be steeper than the north- and south-facing ones. Such a latitudinal categorisation can be no more than tentative at this stage and is useful only insofar as it may aid discussion of the subject and direct research towards particular questions. It is unfortunate that there is no comparable group of studies from the southern hemisphere. Were it to be found that

Figure 4.11: Valley asymmetry showing variation with latitude in azimuth of the steeper valley side. Small diamonds are shown where a publication gives a single value for the azimuth of the steeper valley-side, extended ones are shown where one or more ranges of azimuth are given

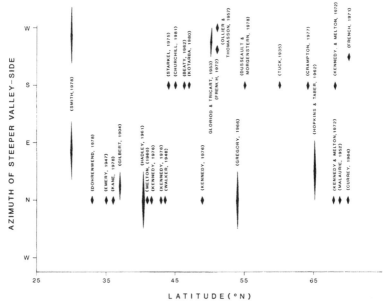

62

the equatorward sequence of steeper valley sides in that hemisphere were south, north, and east and west, it would support the division suggested above.

Even if some consensus on the orientation of the steeper valley side at particular latitudes can be identified, the same cannot be said of the processes thought to be responsible for valley asymmetry. Within the most poleward region there is general agreement that asymmetry results from past or present glacial or periglacial processes. However, the way in which these processes operate to produce asymmetric valleys is a matter of some disagreement. Some authors (e.g. Büdel, 1953; French, 1971) have argued from the evidence of asymmetric valley-floor deposits that differential undercutting of the valley sides by the basal stream results in differences in valley-side steepness. According to this hypothesis, more colluvium is deposited at the foot of the north-facing hillslope because solifluction processes are more active on this valley side. The basal stream is thus forced against, and begins to undercut, the south-facing valley side. Other authors, however, have claimed that the differences in gradient are directly attributable to the effects of hillslope processes. Because processes are more active on one valley side, so one side is steeper. Whether more active processes cause a valley side to steepen or become more gentle and, hence, whether it is the north-facing or the south-facing hillslope that is more active is a subject of disagreement. Malaurie (1952), Starkel (1975), Crampton (1977) and Kotarba (1980) are among those authors who have claimed that more rapid freeze-thaw on the south-facing valley side results in its steepening, whereas Tuck (1935) and Churchill (1981) have argued for more rapid erosion of the north-facing hillslope which, as a result, flattens. It is tempting to suggest that authors have argued their cases not so much from the evidence available but from preconceived notions about the effects of hillslope processes. Those authors who have believed that the operation of processes leads to hillslope decline have argued that, *ergo*, processes are more active on the gentler valley side.

The majority of studies undertaken in the higher latitudes have identified the south-facing valley side as steeper. Kennedy and Melton

63

(1972), however, identified steeper gradients on both north- and south-facing valley sides in permafrost areas of the Northwest Territories, Canada. They argued that asymmetry results from the effects of both microclimate and relief such that in more severe climates of low relief the north-facing valley sides are steeper, but that in the zone of milder and deeper valleys it is south-facing hillslopes that are steeper. The importance of this study lies in its recognition of the fact that asymmetry may have a variety of causes, some of which are climatically controlled. In any environment asymmetry may be the outcome of a number of simultaneously operating (and possibly opposing) processes. Other studies of asymmetry undertaken in similar environments have not attempted to isolate the various factors affecting asymmetry so that it is impossible to determine the extent to which Kennedy and Melton's finding is representative of more than local conditions.

Away from areas affected by glacial and periglacial processes studies of valley asymmetry have been less common. In mid-latitude areas where the north-facing valley side has been found to be steeper (Bass, 1929; Hack and Goodlett, 1960; Emery, 1947; Hadley, 1961; Dohrenwend, 1978; Kane, 1978), this has usually been attributed to climatically-induced differences in vegetation. The south-facing valley side is drier and has a poorer vegetation cover. Hence it is more rapidly eroded. This has been thought to cause the south-facing hillslope to decline more rapidly (Bass, Emery, Kane) or to result in more colluvium on the northern side of the valley floor causing the basal stream preferentially to undercut the north-facing valley side (Dohrenwend).

For the sub-tropics, where the solar zenith is more nearly vertical, Smith (1978) has argued that weathering will be enhanced on east- and west-facing hillslopes because these hillslopes remain in shadow longest, are more likely to receive and retain moisture but still experience rapid diurnal heating and cooling. Smith has found east-west valley sides in the Beni Abbes region of Algeria to be steeper than those of other orientations but has argued that these differences have not been created under present climatic conditions. Rather these conditions maintain an asymmetry inherited from a moister climate.

Other Aspect-Related Differences of Hillslope Form

Although gradient is only one attribute of hillslope form, it has dominated studies of valley asymmetry. Carson and Kirkby (1972, p.387) have pointed to the lack of attention paid to asymmetry of overall hillslope geometry. Churchill (1981) examined differences in 16 morphological attributes between north- and south-facing hillslopes in South Dakota and obtained statistically significant differences in five of them (Table 4.1), three of which do not measure steepness directly. However, because of the geometry of

Table 4.1: Differences between morphological attributes of north- and south-facing hillslopes

Hillslope attribute	Mean value for north-facing hillslopes	Mean value for south-facing hillslopes	F value
Number of convexities	0.5	0.1	3.46
Number of concavities	0.4	0.3	.23
Number of segments	6.4	4.8	6.38 †
Total number, all form units	7.2	5.3	10.01 †
Proportion, convexities	6.9	3.2	1.74
Proportion, concavities	5.9	5.9	.00
Proportion, segments	87.2	90.0	.78
Maximum gradient (degrees)	66.9	76.8	11.56 †
Relative length, maximum segment	14.7	25.6	1.71
Maximum convex curvature (degrees/100 feet)	-74.0	-46.0	.64
Relative length, maximum convexity	6.1	3.2	1.18
Maximum concave curvature (degrees/100 feet)	113.4	170.6	.78
Relative length, maximum concavity	5.1	5.9	.09
Mean gradient (degrees)	47.0	61.1	34.59 †
Profile length (feet)	52.4	44.8	4.18 †
Profile relief (feet)	37.0	37.4	.01

† $F > F_{.05}$

After Churchill (1981, p.379), Annal. Ass. Am. Geogr.,71. *Reproduced by permission*

the hillslopes studied, profile length and, hence, number of form units and number of segments must all be correlated with mean angle. Another factor which may have contributed to the greater number

65

of form units observed on the north-facing hillslopes is hillslope roughness. Ohmori (1979) also found north-facing mountain hillslopes in Japan to have greater roughness. Because of the correlations of roughness asymmetry with both latitude and altitude, Ohmori attributed rougher north-facing hillslopes to greater Pleistocene periglacial activity on these hillslopes. This finding is consistent with that of Liebling and Scherp (1983) who observed that north- and north-east facing hillslopes in parts of the Nahoni Range, Yukon Territory, Canada are more dissected than hillslopes facing south and south-west. They identified this difference with the longer and slower snowmelt on north- and north-east facing hillslopes which resulted in rivulets of water discharging downslope from each patch and embankment of snow. These rivulets cause the greater degree of dissection of these hillslopes. Madduma Bandara (1974) also noted orientational differences in hillslope form. In part of the central Highlands of Sri Lanka this author found rectilinear hillslopes to have a predominantly north-east orientation which was tentatively attributed to these hillslopes being in the lee of the dominant rain-bearing winds.

Aspect-Related Differences in Hillslope Processes

A major problem in the study of valley asymmetry lies in the inference of processes from forms. Most studies have explained an observed pattern of asymmetry in terms of processes whose operation is uncertain and whose effects are unknown. Compared to the number of studies of asymmetry of valley-side form there are remarkably few to show that microclimatic differences are sufficient to cause changes in the nature or rates of hillslope processes on opposing valley sides. Soons and Rainer (1968) have reported on an experiment in which runoff and sediment yield from plots distributed around a drainage basin are measured in relation to meteorological observations. Churchill (1982) has monitored soil dessication on north- and south- facing valley sides following a rainfall event and observed significant differences. He was able to incorporate these differences into a set of empirically-derived causal relations linking aspect, dessication rate, regolith thickness, infiltration index, drainage density and number of mass movements. Such studies are essential if the extent to which microclimatic differences affect hillslope processes is to be determined. Nevertheless, they beg the question of how far contemporary climatic conditions (and processes) are relevant to the explanation of contemporary asymmetry (see Chapter 9). Both Dohrenwend (1978) and Smith (1978) found the contemporary conditions on the hillslopes they studied to be in accord with the patterns of asymmetry they observed, yet each attributed the asymmetry to processes operating under a former climate. There can be no doubt, however, that research on aspect-related differences in process rates and mechanisms provides the only means whereby the topic of climatically controlled valley asymmetry can be raised above the level of a discussion based upon conjecture.

THE INFLUENCE OF MATERIALS 5

The materials in which sub-aerially formed hillslopes are cut range from hard crystalline rocks to loose friable sands. In most environments there is also a layer of soil and/or regolith forming the uppermost portion of the underlying material. The thickness of this layer varies so that the surface form of a hillslope reflects both the shape of the mass of underlying rock or sediment and the local thickness of the soil/regolith layer. Through their physical, chemical and structural properties these materials affect hillslope processes and thereby influence hillslope form. In examining this influence it is convenient to consider materials in three groups: rocks, unconsolidated deposits and soils. The boundaries of these three groups are imprecise. For the purposes of this chapter, materials are termed rocks if they have undergone diagenic processes or otherwise possess structural features (joints, bedding planes etc.) that may influence their effect on hillslope form. The section on unconsolidated materials is concerned primarily with accumulations of loose rock fragments, but also examines cohesive deposits. The discussion of the influence of soils is concerned wholly with the effects of the surface layer which is taken to include both the weathered material that is undergoing pedogenic processes and the vegetation.

ROCKS

Rocks influence the form of hillslopes in three ways: as a function of their strength, through their influence on hillslope processes, and by virtue of their structural features. In practice, these three mechanisms are interrelated, but they provide a convenient framework within which to examine rock control on hillslope form.

Rock Strength

If it is possible to imagine the initiation of a hillslope then that initiation is the result of some agency that leads to a difference in elevation across a plane. Such an agency might be a downcutting river, a tectonic or isostatic movement, or a fall in sea level. A hillslope is the form that is created by sub-aerial processes acting across the plane. The strength of hillslope materials is important not only because it influences the shape of the sub-aerially created hillslope but also because it limits the size of the initial difference in elevation. Consider the difference in elevation shown in Figure 5.1. Above b there is an inequality in the

Figure 5.1: Horizontal and vertical stresses in a rock mass

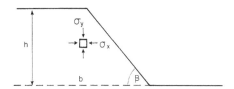

lateral and vertical stresses (σ_x and σ_y) acting on the material. This inequality is maintained by the strength of the material (measured by the unconfined compressive strength q_u). The inequality in σ_x and σ_y increases as h increases so that eventually a critical height h_c is reached beyond which q_u is inadequate to counterbalance the inequality. For a difference in elevation bounded by a vertical plane ($\beta = 90°$), Terzaghi (1962) calculated

$$h_c = q_{u/\gamma} \tag{5.1}$$

where γ is bulk density. For the weakest of hard unweathered rocks, Terzaghi calculated h_c to be 1280 m and instanced intact granite as having a value for h_c several times greater.

In reality differences of elevation do not attain these values and many examples show that

69

rocks fail before these critical heights. Terzaghi argued, therefore, that mechanical defects in rocks, such as joints and faults, restrict differences in elevation to values below h_c. If the agency responsible creates a difference in elevation greater than that which can be resisted by the rock, failure (in the form of mass movement) occurs to lessen the height and/or gradient. This is one way in which a difference in elevation may evolve into a sub-aerially formed hillslope.

Just as rock strength exerts a control in the initial development of a hillslope so it may be expected that, under the action of sub-aerial processes, a relationship will continue to exist between hillslope form and rock strength. Weathering (e.g. opening of joints, chemical alterations in the rock mass) will result in changes in the strength of the rock and the form of the hillslope will adjust accordingly. Selby (1980) has attempted to produce an index of rock mass strength that can be related to the gradient of bare-rock hillslopes. Selby's index is a numerical score obtained from seven properties of rock-cut hillslopes. Each property is measured on a five-point ordinal scale and a numerical rating is assigned to each point of each scale (Table 5.1). The index of mass strength for a particular rock face is given by the sum of its ratings on these scales. The minimum sum of ratings that can be attained is 25 and the maximum is 100. Selby has shown that in the absence of strong structural control (e.g. bedding or joint planes aligned parallel to the hillslope surface) or active denudation (e.g. an undercutting stream), the inclinations of rock-cut hillslopes are strongly correlated with his index of rock strength and he proposed (Selby, 1982b) an envelope relation for what he termed *strength equilibrium* hillslopes. On statistical grounds and with the availability of more data, Moon (1984) and Abrahams and Parsons (1987) have produced modified versions of the strength equilibrium envelope. That proposed by Abrahams and Parsons (Figure 5.2) was based upon all the then available data on strength equilibrium hillslopes and represents the best current estimate for Selby's strength equilibrium envelope.

On the assumption that Selby's index is a sound measure of rock mass strength, the strength

Table 5.1: Geomorphic rock mass strength classification and ratings

Parameter	1 Very strong	2 Strong	3 Moderate	4 Weak	5 Very weak
Intact rock strength (N-type Schmidt Hammer 'R')	100-60 r:20	60-50 r:18	50-40 r:14	40-35 r:10	35-10 r:5
Weathering	unweathered r:10	slightly weathered r:9	moderately weathered r:7	highly weathered r:5	completely weathered r:3
Spacing of joints	> 3 m r:30	3-1 m r:28	1-0.3 m r:21	300-50 mm r:15	<50 mm r:8
Joint orientations	Very favourable. Steep dips into slope, cross joints interlock r:20	Favourable. Moderate dips into slope r:18	Fair. Horizontal dips,or nearly vertical (hard rocks only) r:14	Unfavourable. Moderate dips out of slope r:9	Very unfavourable. Steep dips out of slope r:5
Width of joints	<0.1 mm r:7	0.1-1 mm r:6	1-5 mm r:5	5-20 mm r:4	>20 mm r:2
Continuity of joints	none continuous r:7	few continuous r:6	continuous, no infill r:5	continuous, thin infill r:4	continuous, thick infill r:1
Outflow of groundwater	none r:6	trace r:5	slight <25 1/min/ 10 m² r:4	moderate 25-125 1/min/ 10 m² r:3	great >125 1/min/ 10 m² r:1
Total rating	100-91	90-71	70-51	50-26	<26

After Selby (1980, pp.44-5), Zeit für Geomorph.,24. *Reproduced by permission*

equilibrium envelope provides a means to identify bare-rock hillslopes that have other than a sub-aerial origin. Oberlander (1972) argued that in the central Mojave Desert the rock slopes characteristic of outcrops of coarse-grained quartz monzonite were exposed by stripping of a deep-weathering mantle as the climate became more arid during the late Tertiary and Quaternary.

Figure 5.2: Strength equilibrium envelope for rock slopes (*After Abrahams & Parsons, 1987*, Earth Surface Processes and Landforms,12. Reproduced by permission of John Wiley & Sons Ltd. Copyright 1987)

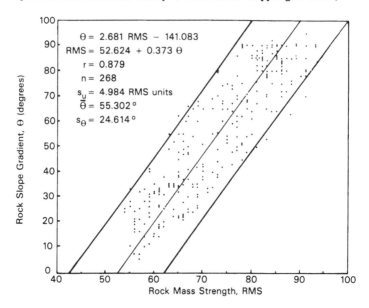

Moreover, he presented qualitative evidence to suggest that while contemporary weathering is reducing boulders now resting on the massive outcrops, it is having little effect on the structurally controlled surfaces of these out-crops. Analysis of a sample of such rock slopes (Figure 5.3) shows that whereas the majority of those inclined at greater than 40° plot within the strength equilibrium envelope, all those at gradients less than this figure fall outside of it. Inasmuch as the maximum gradient of soil-covered hillslopes is approximately 40°, this analysis of rock strength supports Oberlander's thesis that low-angle rock slopes in the Mojave Desert had a sub-surface origin and that they have been minimally affected by weathering since their exhumation.

Equally important are the implications of rock strength for hillslope evolution. Moon and Selby (1983) have argued that if hillslopes remain in strength equilibrium the manner of evolutionary changes must be controlled by changes in rock strength. Specifically, they have contested the view that parallel retreat is associated with free

Figure 5.3: Strength equilibrium relations for rock slopes in the Mojave Desert (*Abrahams & Parsons, unpublished*)

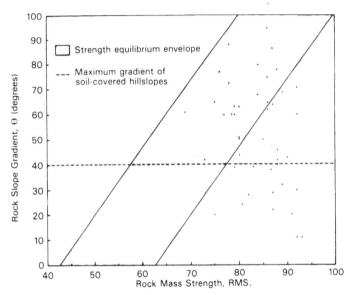

faces on hillslopes. Rock-cut sections of hillslopes will retreat parallel to themselves only so long as there is neither spatial variability nor long-term change in the strength of the rock mass.

Rock Structure

As Terzaghi (1962) has argued, the strength of unweathered rock depends to a large degree on the weakness conferred upon it by its structure. The nature, frequency, continuity and orientation of structural planes within a rock mass all contribute to the manner in which rock structure affects strength. Hensher (1987) has listed 10 types of structural discontinuity that may be present in rock masses (Table 5.2). According to their origin, structural discontinuities in rock masses vary in their surface morphology, frequency and continuity. All of these properties influence

73

Table 5.2: Types of structural discontinuity found in rocks

Occurrence	Type	Physical characteristics
Common to all rocks	Tectonic joints	Persistent fractures resulting from tectonic stresses. Joints often occur as related groups or 'sets'. Joint systems of conjugate sets may be explained in terms of regional stress fields.
	Faults	Fractures along which displacement has occurred. Any scale from millimetres to hundreds of kilometres. Often associated with zones of sheared rock.
	Sheeting joints	Rough, often widely spaced fractures; parallel to the ground surface; formed under tension as a result of unloading.
	Lithological boundaries	Boundaries between different rock types. May be of any angle, shape, and complexity according to geological history.
Sedimentary rocks	Bedding planes/ bedding plane joints	Parallel to original deposition surface and marking a hiatus in deposition. Usually almost horizontal in unfolded rocks.
	Shaley cleavage	Close, parallel discontinuities formed in mudstone during diagenesis and resulting in fissility.
	Random fissures	Common in recent sediments probably due to shrinkage and minor shearing during consolidation. Not extensive but important mass features.
Igneous rocks	Cooling joints	Systematic sets of hexagonal joints perpendicular to cooling surfaces are common in lavas and sills. Larger intrusions typified by doming joints and cross joints.
Metamorphic rocks	Slaty cleavage	Closely spaced, parallel and persistent planar integral discontinuities in fine-grained strong rock.
	Schistosity	Crenulate or wavy foliation with parallel alignment of minerals in coarser-grained rocks.

After Hensher (1987, pp.147-8), Slope Stability, *M.G.Anderson & K.S.Richards eds. Reproduced by permission of John Wiley & Sons Ltd.* Copyright 1987

the degree to which any particular discontinuity type may determine hillslope form.   In granites, widely spaced, curved sheeting joints commonly give rise to domical tors and inselbergs (Figure 5.4), whereas more closely spaced, rectangular, tectonic joints result in tors and inselbergs that typically have a more castellated form (Figure 5.5).

The   orientation   of   discontinuities   is important in their effect on hillslope processes and forms.   Where discontinuities are inclined against the   gradient   of   a   hillslope   they strengthen the rock mass but where they are inclined in the direction of the hillslope they promote instability (Figure 5.6).   It is worth

74

noting that many of the best examples of four-element hillslopes are found in rocks where the structural planes are horizontal.

Figure 5.4: A domical granite tor formed where joints are curved and widely spaced

Structural discontinuities within rock masses are not only planes of weakness in themselves, but also mark the boundaries between stronger units of rock (i.e. those with few or no planes of weakness within them). Accordingly, it is along these planes that processes of erosion are most frequently temporarily halted. This is the origin

Figure 5.5: A castellated granite tor formed where joints are rectangular and closely spaced

Figure 5.6: Orientations of rock discontinuities in relation to hillslope gradient: (a) joints are inclined against hillslope gradient and enhance rock strength, (b) joints are horizontal and have near neutral effect on rock strength. (c) joints dip gently out of the hillslope and moderately weaken rock strength, (d) joints dip steeply out of the hillslope and greatly weaken rock strength

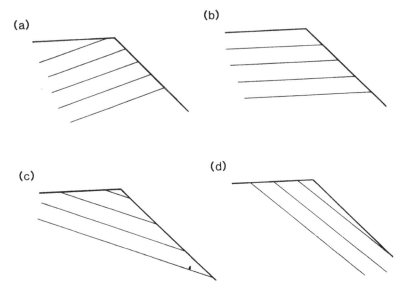

of many so-called "structurally controlled" hillslopes (Figure 5.7). As elsewhere, the form of these hillslopes is a response to the operation of sub-aerial processes. These processes yield hillslopes whose forms reflect structural discontinuities because the discontinuities mark variations in the resistance to erosion in the rock mass.

Rock structure includes not only the disposition and attitude of planes of weakness within lithologically uniform rock units but also the relationships among lithologically different rock units. Young (1972, p.215) argued that the super-position of strata of different lithologies has a greater effect on hillslope form than the structural properties of individual lithological units. The most important relationship is that between a resistant rock overlying a less resistant one. The weakness of the underlying rock controls retreat of the more resistant caprock, the weight of which in some instances is sufficient to cause extrusion of the lower

Figure 5.7: A "structurally controlled" hillslope

lithology and a consequent downward curvature to the margin of the upper rock unit (see Figure 5.8), a process termed *cambering*.

Figure 5.8: Cambering in a resistant caprock

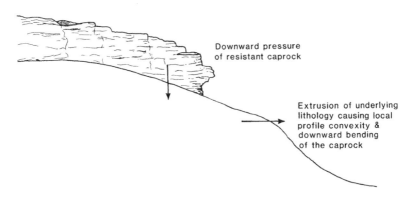

Downward pressure of resistant caprock

Extrusion of underlying lithology causing local profile convexity & downward bending of the caprock

Finally, it should not be forgotten that rocks attain their structures as a result of tectonic processes. Although some such processes operate prior to the formation of sub-aerial

77

hillslopes others may operate simultaneously with denudational processes and thus influence the operation of these processes in a more direct way. It was as an attempt to use hillslope form as a means to infer tectonic processes that Penck (1924) undertook his analysis of hillslope forms. Recent study of tectonic geomorphology has given little attention to the effects of tectonic processes on hillslope form other than to draw attention to the importance of earthquakes and landslides in tectonically active areas (Simonett, 1967; Crozier, Gage, Pettinga, Selby and Wasson, 1982). In consequence, it can be expected that in such areas hillslope forms will be those associated with a high incidence of mass movement processes (see chapter 7).

The Effects of Rock Type on Hillslope Processes

Unfortunately, there are no studies of the effects of rock type on hillslope processes comparable to the surveys conducted by Young (1974b) and Saunders and Young (1983) for the effects of climate, although the data used in these surveys may be amenable to classification by lithology. It is commonly acknowledged that solutional processes are more effective on limestones and this is borne out in the survey by Saunders and Young (p.486, Figure 3). Within a more restricted environment, Waylen (1979) has obtained estimates of ground surface lowering by solution for a variety of rock types (Table 5.3).

In the absence of observational data, how far rocks affect hillslope processes is a matter of speculation. Certainly properties such as mineralogical composition and permeability can be expected to affect susceptibility to particular hillslope processes. For example, Engelen (1973) observed differences between the processes operating on different parts of the Chadron and Brule formations in South Dakota. On the clayey parts soil flowage appeared dominant whereas the sandy mud- and siltstones were subject to rilling. However, in general, the magnitude of such variations in dominant processes among rock types and their effects on hillslope form remain unknown. Ahnert (1976a) has approached this problem by incorporating data on variable rock resistance into a process-response model of hillslope evolution.

Table 5.3: Solutional denudation on some selected lithologies

Lithology	Location	Solutional denudation (mm/1000 yr)	Source
Old Red Sandsone	Mendip	1.6	Waylen (1976)
Clay-with-flints	S.E. Deven	14.0	Walling (1971)
Keuper Marl	S.E. Devon	22.8	Walling (1971)
Upper Greensand	S.E. Devon	16.8	Walling (1971)
Keuper Marl-Upper Greensand	S.E. Devon	33.8	Walling (1971)
Keuper Marl-Upper Greensand	S.E. Devon	27.6	Walling (1971)
Carboniferous Limestone	Mendip	40	Corbel (1957)
Carboniferous Limestone	Mendip	38-45	Ford (1963)
Carboniferous Limestone	Mendip	50-102	Drew (1967)
Carboniferous Limestone	Mendip	22.8	Newson (1970)
Carboniferous Limestone	Mendip	81	Atkinson (1971)
Silurian Greywackes	Montgomeryshire	2.1	Oxley (1974)
Silurian Greywackes	Montgomeryshire	1.6	Oxley (1974)
Middle Jurassic-Liassic clays and shales	North York Moors	20.2	Imeson (1974)
Chalk	Yorkshire Wolds	22.6	Imeson (1974)
Chalky Boulder Clay	Holderness, Yorkshire	49.1	Imeson (1974)
Precambrian Granite, Palaeozoic and Mesozoic sediments, Tertiary and Pleistocene deposits	N.E. Wind River Range, Wyoming	6.8	Hembree and Rainwater (1961)
	S.W. Wind River Range, Wyoming	3.6	Hembree and Rainwater (1961)
Precambrian Quartzite	New Mexico	0.8-1.5	Miller (1961)
Precambrian Granite	New Mexico	0.5-5.5	Miller (1961)
Carboniferous Sandstone	New Mexico	5.8-21.5	Miller (1961)
Glacial Till	New Hampshire	32	Johnson et al. (1968)
Micaceous Schist	Maryland	2.7	Cleaves et al. (1970)

After Waylen (1979, p.177). Earth Surface Processes and Landforms,4. *Reproduced by permission John Wiley & Sons Ltd. Copyright 1979*

Hillslope Form and Rock Type

Because of the paucity of data on the variation of hillslope form with rock type, the results reported in this section are more piecemeal illustrations of the kind of information that might be collected rather than an indication of how far hillslope form varies with rock type. No studies have attempted to control for the effects of other variables affecting hillslope form. In practice, it is probably unrealistic to look for such control. Any conclusion that might some day emerge concerning the influence of rock type on hillslope form is more likely to appear as a consensus from a group of disparate studies than as the result of one or more controlled experiments.

Debris Slopes
In arid environments hillslopes are commonly mantled with a layer of loose rock fragments.

79

Such hillslopes have been variously termed *boulder-controlled, debris-mantled* or *debris-covered*, but may be more simply termed *debris slopes*. The nature of the relationship between gradient and size of rock fragments on these hillslopes has been the subject of debate since the work of Lawson (1915) and Bryan (1925). Abrahams *et al.* (1985) have shown the relationship to be a function of rock type. Although these authors identified strong correlations between gradient and mean size of surface rock fragments, they demonstrated that on hillslopes of the same gradient a different mean size of rock fragment may be found depending upon the weathering characteristics of the underlying lithology (Figure 5.9).

Figure 5.9: Gradient-particle size relations for debris slopes underlain by gneiss G, latitic porphyry LP, and fanglomerate F (*After Abrahams, Parsons & Hirsch, 1985,* Hillslope gradient - particle size relations: evidence for the formation of debris slopes by hydraulic processes, J. Geol., 93, 347-57, Copyright 1985 by the University of Chicago. *Reproduced by permission of the University of Chicago Press*)

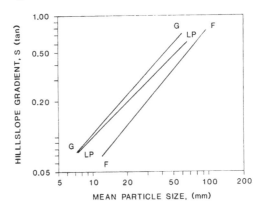

Piedmont Junctions

The sharpness of the junction between the hillslopes and piedmonts of arid areas also varies with lithology, and particularly with the manner of rock weathering. On lithologies that weather by granular disintegration, such that the debris mantling hillslopes and piedmonts has a bimodal size distribution (Figure 5.10a), piedmont junctions are abrupt and angular (Figure 5.11a). In contrast, on those lithologies that weather by fragmentation to yield debris with a multimodal distribution (Figure 5.10b) the piedmont junction is wide and smoothly curved (Figure 5.11b). In some instances sharp piedmont junctions mark a break in the dominant control of gradient. On the hillslopes it is the weathering characteristics of the rock that are the dominant control whereas on the piedmonts transportational processes are the chief determinant of gradient. Elsewhere, however, the gradients of both hillslope and piedmont are determined by denudational capacities and the difference in form is a response to the effects of the different patterns of rock breakdown on hillslope processes. On lithologies characterised by a bimodal size distribution of weathering products, overland flow is concentrated into smooth sandy areas between boulders. For lithologies whose weathering products have a more uniform size distribution overland flow is dispersed. Where flow is concentrated it deepens downslope rapidly and is matched by a rapid decrease of gradient. Conversely, the dispersed flow deepens downslope only slowly and the reduction in gradient is correspondingly slow.

Figure 5.10: Size distributions of two debris mantles
A. From a rock which weathers by granular disintegration
B. From a rock which weathers by fragmentation
(*After Parsons & Abrahams, 1987*, Gradient-particle size relations on quartz monzonite debris slopes in the Mojave Desert, J. Geol., 95, 423-32 Copyright 1987 by the University of Chicago. *Reproduced by permission of the University of Chicago Press*)

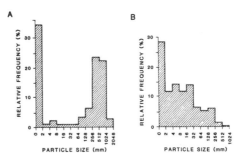

81

Figure 5.11: Contrast in piedmont junction concavity
  A. Abrupt, angular concavity
  B. Wide, smoothly curved concavity

Profile Curvature
Several authors have examined the relative length of convex, concave and rectilinear sections in hillslope profiles and related these proportions to lithology. In a study of several rock types in central Belgium, Fourneau (1960) found that convexities occupy the greater part of profiles on sandstones, but that this proportion is reduced to about 50% on limestones and less than 50% on shales. Christofoletti and Tavares (1976) and Kumar (1981) used an index of profile curvature I determined from the length of convex and concave portions of a profile (Lx and Lv, respectively) and the angular change over convex and concave portions (Ix and Iv, respectively) so that

$$I = (Lx/Ix)/(Lv/Iv)$$

For profiles measured on crystalline basement rocks in Minas Gerais, Christofoletti and Tavares found indices ranging from 0.59 to 2.51 indicating a wide variety of profile forms. Kumar, in a smaller sample of profiles on pre-Cambrian quartzites in Bihar, likewise identified both dominantly convex and dominantly concave profiles ($0.42 < I < 1.55$).

In a more comprehensive analysis of the morphology of chalk hillslopes, Clark (1965) distinguished between convexo-concave (Cx-Cv) hillslopes, those that had a rectilinear midslope segment (Cx-R-Cv) and more complex profiles. Of the sample of 95 hillslopes, 32% fell into the Cx-Cv category and for these hillslopes the convexity was found to occupy an average of 63% of profile length. Although the convexity reduces to an average of 37% of profile length on Cx-R-Cv hillslopes (47% of the sample), the rectilinear segment occupies 41% so that concavity is still the smallest part of the profile.

Simple analysis of hillslope profile curvature is a useful approach to the issue of rock control on hillslope form. However, a much larger set of data than is presently available will be needed before any significant conclusions can be drawn.

Microrelief
There is some evidence to suggest that lithology affects hillslope microrelief. In central Belgium, Fourneau (1960) found convexities on limestone to be smoother than those on sandstone, but Lambert (1961) showed the converse to be true

on hillslopes that he studied in the Condroz area of southern Belgium. Knill (1982) undertook a study of microrelief on several rock types and identified differences in magnitude and shape of microrelief as a function of the underlying lithology.

The Influence of Rock Type on Hillslope Evolution

Several authors have speculated on the manner of hillslope evolution in relation to rock type. Savigear (1960) claimed that in West Africa hillslopes evolve by decline where they are formed in non-massive rock but that on massive rocks evolution is by slope replacement. He argued that on lithologies with closely spaced interstices waste production is so fast, relative to the rate of waste removal, that the bedrock becomes blanketed. On massive rocks the initial rate of production of waste blocks is very slow by comparison with that of their subsequent break up and removal. Kotarba (1986) believed that rockwalls, straight debris-covered hillslopes, and convex hillslopes represented an evolutionary sequence for granite hillslopes in semi-arid Mongolia. Ollier and Tuddenham (1962) contrasted the evolution of hillslopes beneath a duricrust cap with that of those from which the cap had been eroded. So long as it is present, the duricrust controls the retreat of the hillslope as a whole. Beneath the backwearing cap, profile form remains the same as the hillslope retreats parallel to itself and the pediment at its base is extended. Once the duricrust is removed gradients on the underlying softer rocks are rapidly reduced. Pallister (1956b) used a similar argument for hillslope development in Buganda. The problem with studies of this type lies in knowing that spatial differences of hillslope form can be fitted into an evolutionary sequence. Pain (1986) used evidence for the evolution of drainage in eastern New South Wales to place his hillslopes in a relative chronology. In his study area a sandstone caprock overlies shale, a situation analogous to that of a duricrust caprock. Pain, likewise argued for parallel retreat and slope replacement while the caprock remains in place and for slope replacement and decline once it has been eliminated (Figure 5.12).

The most satisfactory of the studies that

Figure 5.12: Effect of a caprock on hillslope evolution (*After Pain, 1986*, Catena,13)

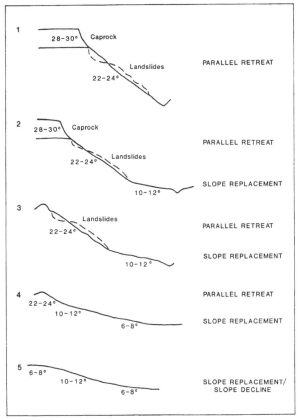

have claimed an evolutionary significance for a set of profile measurements have been those that have been made along abandoned coastal cliffs. An indication of the effect of lithology on the evolution of such hillslopes can be obtained by comparing the study undertaken by Savigear (1952), on an Old Red Sandstone cliffline in South Wales, with that of Hutchinson and Gostelow (1976), on a London Clay cliff in Essex. In both studies an accumulation zone at the foot of the abandoned cliff was identified following cessation of marine erosion. Savigear argued that this represented a condition of impeded removal and his evidence showed that under this condition gradient declined in contrast to parallel retreat where removal was

85

unimpeded. Hutchinson and Gostelow also presented evidence for slope decline and retreat. Although the upper part of the cliff is capped by sands and is thought to have undergone parallel retreat, the lower part is cut in London Clay and is shown to have declined in gradient from 17.1° at 10,000 yrs B.P. to 10.8° at the present time. They further argued that the gradient of the accumulation zone (7.7°) is close to the ultimate angle of stability of London Clay against landsliding and may be considered to anticipate the eventual gradient of the entire cliff. Savigear did not speculate on the ultimate form of the cliff that he studied nor did he present geotechnical evidence of the type used by Hutchinson and Gostelow. Nevertheless, it is not unreasonable to suggest that despite the evidence of slope decline on two very different lithologies in response to similar basal conditions, the lithological differences may, in the end, determine the gradient of the fully degraded cliff.

The special case of calcareous rocks, on which hillslope evolution is dominated by solutional processes, is too large to be examined here. Some consideration is given to the effects of solution on hillslope evolution in chapter 7; otherwise readers are referred to more specialised texts on karst landforms (e.g. Jennings, 1985; Trudgill, 1985).

UNCONSOLIDATED DEPOSITS

Hillslopes developed in unconsolidated deposits lend themselves more easily to an analysis of their form in geomechanical terms than do those cut in rock. Thus, for example, Lohnes and Handy (1968), using the equation

$$H_c = \frac{4c}{\gamma[\cos\varphi - 2\cos^2(45° + \varphi/2)\tan\varphi]} - y$$

in which c is cohesion, φ is the angle of internal friction, y is the depth of any tension crack and other terms are as in equation 5.1, were able to demonstrate a close relationship between the observed heights of cliffs in loess within their study area and the predicted maximum height H_c. Likewise, Matsukura, Hayashida and Maekado (1984) obtained close agreement between gradients

86

measured on ignimbrite in South Kyushu and those predicted from stability analyses. These results stand in contrast to the considerable overestimate of predicted cliff height in rock obtained by Terzaghi (1962) using a similar equation. Indeed, the equation developed by Coulomb (1776) for shear strength s:

$$s = c + \sigma.\tan\varphi$$

provides a convenient framework for the analysis of hillslopes in unconsolidated deposits.

The role of the angle of internal friction, φ

Probably nowhere else in the study of hillslopes has there been closer attention to the relationship of hillslope form to materials than in the study of the gradients of talus slopes. Ostensibly the problem is a simple one. Rock fragments, released by weathering, fall from a near vertical cliff and accumulate at its base. What determines the gradient (angle of repose) of the accumulating pile of rock fragments? Rankine (1857) defined the tangent of the angle of repose as equal to the coefficient of friction determined by the ratio of normal pressure to resistance to sliding. Van Burkalow (1945) accepted that the angle of repose represents the steepest gradient that can be maintained by a pile of loose material, and Skempton (1945) suggested an approximate equality of the angle of repose and the angle of internal friction. In her laboratory study, Van Burkalow found the angle of repose to vary (1) inversely with the size of fragments in perfectly sorted materials but directly in those imperfectly sorted; (2) inversely with density of fragments; (3) directly with fragment angularity, roughness and degree of compaction; and (4) inversely with height of fall onto free cones.

More recently, it has been disputed both that the angle of repose of rock fragments piled up in such a manner is equal to the angle of internal friction and that this angle is the steepest that a pile of loose material can maintain. Metcalf (1966) reported experiments on loosely packed fragments (such as would be found in a talus slope) and on closely packed fragments (which had a 10% lower void ratio) and claimed that the angle of repose better matched the angle of internal friction of the closely packed fragments. Allen

(1969) pointed to the fact that loose material could be accumulated in a pile up to a maximum gradient termed the *angle of initial yield* φ_i. Once this critical angle is exceeded the surface becomes unstable and an avalanche of particles flows away. The gradient of the surface left behind, which is commonly regarded as the angle of repose, Allen termed the *residual angle after shearing* φ_r. Accordingly,

$$\varphi_i = \varphi_r + \Delta\varphi,$$

where $\Delta\varphi$ is simply $\varphi_i - \varphi_r$. Allen argued that experimental evidence shows that for a given granular material φ_r varies little. On the other hand, φ_i is very variable, depending upon the concentration of the rock fragments as they accumulate. Consequently, $\Delta\varphi$ also varies, and empirical data shows it to lie in the range 0° to 14°. Allen developed a theoretical model for spheroids to show that φ_i depends on their concentration which, in turn, depends on their axial ratio.

Observations on natural talus slopes led Chandler (1973) to conclude that their gradients were less than φ_r, a view that was subsequently disputed by Carson (1977). It was Chandler's view that a model of accumulation of loose fragments is inappropriate for most natural talus slopes, on which degradational processes such as creep and fluvial action are more active under present-day conditions than the supply of rockfall material. These degradational processes contribute to determine the gradients of talus slopes and are insignificant only in cases where the supply rate of rock fragments is very high or where the talus slope is being rapidly eroded at its base so that new avalanches of particles are very frequent. According to Chandler, it is only in these situations that the gradient of the talus slope will be close to φ_r.

Many talus slopes exhibit basal concavities (e.g. Rapp, 1959; Caine, 1969; Statham, 1976; East and Gillieson, 1979; Ballantyne and Eckford, 1984). This concavity has traditionally been used as evidence for the operation of processes other than rockfall. Caine attributed the concavity to slush avalanching. However, Statham demonstrated that a basal concavity can occur when rockfall is the only process operative. The critical difference appears to be that under rockfall the basal concavity becomes less as time passes (and the supplying rockwall diminishes in height) whereas slush avalanching leads to a growth in the size of the concavity. There is no basis for resolving this issue and, indeed, it is quite possible that in some environments both processes operate.

The effect of cohesion

For unconsolidated materials that are also cohesive, the situation is less clear. The simple rockfall model of accumulation is not appropriate, and basal erosion in cohesive, unconsolidated materials does not automatically result in particle avalanching, but may yield a vertical hillslope so long as its height is less than H_c. Cohesion results from a number of forces such as the electrostatic and electromagnetic forces that operate between clay-sized particles. An important contributor to cohesion is water. Because water molecules behave as dipoles in the presence of an electric field, they adopt preferred orientations and thus contribute to the cohesive forces between particles.

In contrast to the numerous studies of talus slopes, there have been comparatively few analyses of gradients on cohesive unconsolidated deposits. Those by Lohnes and Handy (1968), on friable loess, and Matsukura *et al.* (1984), on ignimbrite

Figure 5.13: Calculated versus observed maximum heights of 12 loess cuts. (*After Lohnes & Handy, 1868,* Slope angles in friable loess, J. Geol.,76, 247-58 Copyright 1968 by the University of Chicago. *Reproduced by permission of the University of Chicago Press*)

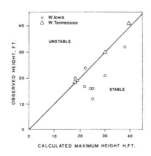

are notable exceptions. In their study Lohnes and Handy determined the shear strength of loess taken from a number of sites and used this to calculate the maximum attainable hillslope heights for the gradients present at the sample sites. Comparing these predictions to field measurements they obtained good agreement (Figure 5.13) and showed many hillslopes to be close to their maximum theoretically attainable height.

Elsewhere (Skempton and DeLory, 1957), it has been argued that the morphology of hillslopes in fissured clays can be understood only if it is assumed that the clay behaves as though it were cohesionless. This result points to the difference between materials that, for the purposes of this chapter, have been termed rocks and those that have been termed unconsolidated deposits.

Evolution of hillslopes in unconsolidated deposits

Probably because the rate of change is comparatively rapid on hillslopes developed in unconsolidated deposits, such hillslopes have attracted a number of studies of their evolution. However, even here the rates are not sufficient for change to be observed directly, and recourse has generally been made to the same space-time substitution as has been the case for rock-cut hillslopes. The significant difference is that because change is comparatively fast, absolute ages for particular hillslopes are relatively easy to obtain. Studies undertaken on both natural hillslopes (Brunsden and Kesel, 1973; Sterr, 1985) and on artificial ones (Goodman and Haigh, 1981; Haigh, 1985) have all identified reduction in overall hillslope gradient through time. None the less, the growth of low-angle, colluvial footslopes and rounding of crests appear to be the dominant mechanisms. Slope replacement rather than slope decline appears characteristic. The results of these studies provide useful insights into the manner of evolution of hillslopes developed on unconsolidated materials but they cannot be taken as indicative of the evolution, over a longer period of time, of rock-cut hillslopes. Specifically, because morphological change is rapid, weathering and processes operating in the soil layer are wholly different from the conditions that obtain on rock-cut hillslopes. In addition, the low-angle footslope is commonly shown as being fully aggradational (Figure 5.14) whereas on rock-cut hillslopes this is seldom the case.

Figure 5.14: Evolution of a hillslope developed in unconsolidated deposits (*After Brunsden & Kesel, 1973*, Slope development on a Mississippi River bluff in historic time, J. Geol.,81, 576-97 Copyright by the University of Chicago. *Reproduced by permission of the University of Chicago Press*)

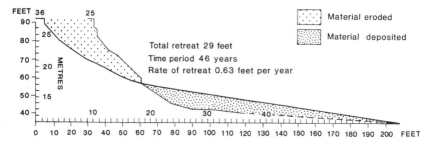

Studies of hillslopes on unconsolidated materials have raised the question of the validity of space-time substitution. Brunsden and Kesel (1973) have pointed out that any set of hillslope profiles will exhibit variation and hence there is no basis for assuming that the variation that is observed when the profiles are arranged according to age is due wholly to evolutionary change. Indeed, they argued that for a set of profiles measured under uniform conditions of basal erosion the variation was due to random fluctuation about a mean profile form. Welch (1970), in studying hillslopes on recessional moraines of the Athabasca Glacier, had reached a similar conclusion. He claimed that there was initial rapid change on exposure of a moraine as it adjusted to new conditions but that after 35 years no further significant morphological change took place.

THE SOIL LAYER

Inasmuch as many hillslopes are mantled by a soil layer, it is upon this layer that hillslope processes operate. The ability of this layer to

91

withstand denudational processes must therefore be important for hillslope form. Melton (1957) found a significant correlation between wet soil strength and valley-side gradient. Souchez (1966) identified an inverse relationship between maximum hillslope gradient and the ratio of the percentage of soil particles finer than 100 μm to the Trask sorting index. In a discussion of soils developed on a variety of rocks in the Oxford region, Chorley (1959, 1964b) attempted to relate their strengths as measured by a cone penetrometer to measures of moisture content, density and grain size. He argued that it may be variations in soil strength that are responsible for differences in relief found upon different lithologies rather than differences in the rocks *per se*.

Shallow landslides, in which the failure plane is approximately planar and parallel to the ground surface, can take place wholly within the soil layer. This type of landsliding is probably the most common form of instability process on natural hillslopes (Carson and Kirkby, 1972, p.152). Accordingly, in environments where this process dominates, it may be the ability of the soil layer to withstand sliding that determines hillslope gradient. A sequence of rock weathering, soil formation, shallow sliding, and renewed exposure of the underlying rock may continue until such time as the gradient of the rock surface is reduced to the point at which the subsequently weathered top layer can withstand shallow sliding. However, the process of weathering is continuous so that the ability of the soil layer to withstand shallow sliding will vary in response to changes in c and φ. In general, as weathering proceeds, voids within the soil become smaller and the chances of positive pore pressures increase. It was this type of reasoning that Carson and Petley (1970) used to explain rectilinear hillslopes at a variety of gradients in Exmoor and the southern Pennines.

The significance of the soil layer for hillslope development under a wider range of processes has been the subject of study by Dietrich Wilson and Reneau, (1986). Hollows in the bedrock, filled with soil as a result of soil creep and wash processes, are subject to periodic flushing by shallow landsliding. Size and shape of soil-filled hollows were found to be related to φ properties of the fill.

Vegetation

Although, strictly speaking, vegetation is not part of the soil layer, it is unreasonable to exclude it from a consideration of the significance of the soil layer for hillslope form. In his study of shearing resistance of soils, Chorley (1964b) included the weight of root matter as one of the independent variables. Furthermore, there is evidence to show that changes in vegetation affect the susceptibility of a soil to shallow landsliding (DeGraff, 1979). The role of vegetation in hillslope stability was the subject of an extensive review by Greenway (1987), who showed that this role could be subdivided into its effects on hillslope hydrology and its influence on the mechanical strength of the soil layer. Hillslope hydrology is affected by the influence of vegetation on interception, infiltration and transpiration. Marsh and Koerner (1972) have demonstrated how moss, and particularly the formation of moss lobes, affects hillslope microrelief, an important factor in determining infiltration. Through its effect on transpiration

Figure 5.15: Effect of roots on soil shear strength (*After Gray & Leiser, 1982,* , Biotechnical Slope Protection and Erosion Control, Copyright Van Nostrand Reinhold. *Reproduced by permission*)

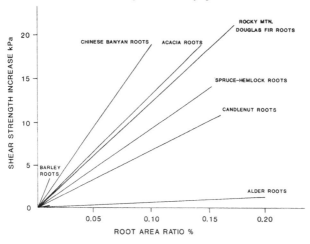

vegetation influences both sediment yield from hillslopes and the balance of processes, as

93

demonstrated by Haigh (1979, 1980). The mechanical influences of vegetation are primarily manifested in the modifications to the shear strength of soil induced by the presence of roots as shown in Figure 5.15.

The effects of vegetation on hillslope form *per se* are less apparent. It has been suggested by Haigh (1977) that the manner of evolution of spoil-bank hillslopes varies with vegetation cover. Vegetated hillslopes display a tendency to parallel retreat and little profile modification, whereas evolution on unvegetated hillslopes is more complex and involves substantial changes in profile form. In addition, it might be inferred that vegetation, through its influence on shear strength of the soil, will affect threshold gradients of mass movements of the type discussed by Carson (1969), thereby influencing the gradients of rectilinear hillslope segments (Carson and Petley, 1970).

LOCAL CONTROLS OF HILLSLOPE FORM 6

Climate and materials exert control on hillslope form at a broad scale. Within an area of uniform climate or material there is variation in hillslope form as shown, for example in Figure 6.1. It is with such variation and its determinants that this chapter is concerned.

Figure 6.1: Effect on hillslope form of varying basal conditions. The lefthand part of the hillslope is being undercut by the basal stream whereas the righthand part is not

RELIEF AND SCALE

A fundamental question in geomorphology, not just in hillslope studies, is that of whether dimension affects morphology. Are landforms scale-free or scale-bound? In a wide-ranging discussion of size and scale in geomorphology, Church and Mark (1980) touched briefly upon hillslopes in presenting Melton's (1958) equation showing the relation between valley-side gradient (θ) and drainage density (L) and relief (H):

$$\theta \propto L^{0.25}H^{0.50}$$

They observed that valley-side gradient and drainage density are positively related if relief is constant. This is a necessity on purely geometric grounds. As drainage density increases so valley sides become shorter. In consequence they must be steeper. If drainage density is held constant, valley-side gradient and relief are positively related. This inference points to relief as a control of hillslope form, proposed earlier in a qualitative way by Glock (1932). This author argued that as valleys develop, an original upland surface of low angle is consumed and replaced by a valley-floor surface, also of low angle. Between these two surfaces are the steeper valley sides. In areas of low available relief Glock contended that valley floors develop before the upland surface has been consumed. Hence average gradient over the land surface remains low throughout the period of valley development. In areas of high available relief the upland surface is consumed by the valley sides before substantial valley floor development has taken place. Hence steep valley sides dominate the land surface for a considerable part of the time.

The importance of relief in controlling hillslope gradient has also been investigated for the case of fault scarps. Wallace (1977) identified scarp height as an important variable controlling the rate of degradation of fault scarps and asserted that along a fault scarp of given age the taller segments display steeper gradients than the shorter ones. He argued that the processes of scarp erosion (which he assumed to lead to a reduction in gradient) operate more rapidly on smaller fault scarps. Wallace did not explain why this should be so and it is difficult to reconcile his view with the results of other work on the effects of hillslope size on erosion processes (e.g. Lal, 1984) and considerations of critical height (see chapter 5). Nevertheless, his assertion accords with the theoretical predictions of Nash (1980a) and is supported by Bucknam and Anderson (1979). The latter authors undertook a more quantitative analysis of the relationship between scarp gradient and scarp height and proposed that it could be approximated by a logarithmic curve (Figure 6.2).

Figure 6.2: Scarp height-gradient relations. Dots show observed scarp gradients and scarp heights. Open triangles show observed scarp gradients and scarp heights of the Bonneville shoreline in the same areas. Solid and dashed lines are from regressions of scarp gradients on the logarithm of scarp height (*After Bucknam & Anderson, 1979 Geology,7, 11-14. Reproduced by permission*)

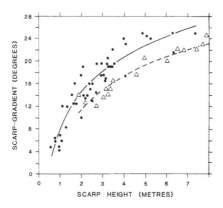

BASAL CONDITION

Variations in the conditions at the base of a hillslope have been the subject of both empirical study and theoretical modelling. Savigear (1952) used temporal variation in the conditions of basal removal along an abandoned cliffline to identify the manner of hillslope development. His results, however, may be interpreted wholly in terms of the effects of varying conditions of basal removal along one hillslope. Savigear recognised conditions of impeded and unimpeded removal of

97

talus from the base of the hillslope. He argued that in the latter case hillslopes are characterised by segments which meet in angular intersections whereas in the former case hillslopes are composed of convex or concave elements which meet without obvious angular discontinuity (Figure 6.3). Similar differences in conditions of basal removal have been considered from a theoretical viewpoint by Bakker and Le Heux (1947, 1950) and Van Dijk and Le Heux (1952). These authors were concerned to model the

Figure 6.3: Changes in hillslope and cliff profiles from faceted to smoothly curved as basal conditions change from unimpeded removal (profiles 1-3) to impeded removal (profiles 4-6) (*After Savigear, 1952*, Inst. Br. Geogr. Trans.,18. *Reproduced by permission*)

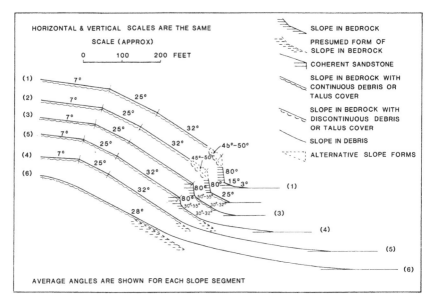

form of the bedrock surface beneath an accumulating waste sheet and not the surface form of the hillslope which they assumed to be rectilinear. As conditions of basal removal change so does the rate of waste accumulation. In consequence, the form of the bedrock surface changes. An important limiting condition is where the accumulation rate is zero (the condition of unimpeded basal removal). In this case the bedrock surface and the surface of the hillslope are coincident. The hillslope that results is a Richter denudation-slope. Examples of hillslopes that conform to these predictions and that are in localities where the conditions of basal removal could well match those required of the model do certainly exist (e.g. Figure 6.4), though this does not of itself validate the model.

Figure 6.4: Richter denudation-slope located where all waste is removed by a basal stream

The effects of more active basal conditions on hillslope form have been considered in a theoretical fashion. Movements of the base of a hillslope, both laterally and vertically, have

been examined. Such movements are generally attributed to lateral and vertical movement by streams that commonly abut the bases of hillslopes. Using a previously developed model (Scheidegger, 1961b), Scheidegger (1960) examined the effects of lateral action of rivers cutting away at the bottom of hillslopes. He concluded that the result of such river action is to change fundamentally the development pattern of a hillslope. The hillslope becomes steeper with the passage of time and asymptotically reaches an inclination determined by the relative effectiveness of river action on the one hand and hillslope processes on the other. From this point onward parallel recession of the hillslope occurs. Scheidegger's study indicates that lateral undercutting of the base of a hillslope has a critical effect on its morphology.

The significance of vertical downcutting by a river at the base of a hillslope has been considered by Armstrong (1982) and Kirkby (1983) within the context of Kirkby's (1971) hillslope process-response model. Armstrong drew attention to two limiting cases. Under the assumption of a constant lower boundary and a fixed hillslope length the model yields a time-independent characteristic hillslope form:

$$y/y_0 \propto \cos(\pi x/2x_1)$$

But, where the lower boundary of the hillslope downcuts at a constant rate, the hillslope adopts a constant form relative to its moving base:

$$(y - y_0) \propto (x_1^2 - x^2)$$

Whereas the characteristic form shows progressive reduction of gradient with time, the constant form retains its steepness. Kirkby (1983) maintained that most hillslopes affected by basal streams would evolve along a path somewhere between these two limiting cases.

Simultaneous lateral and vertical shifts in the position of a basal stream were considered by Hirano (1981) in a model for cross-valley profiles. From this model Hirano identified the ratio of the horizontal and lateral shifts in the position of the river as a critical parameter determining the form of cross-valley profiles. He subsequently used the effects of the ratio on cross-valley profile form to identify differences in the history of lateral and vertical downcutting along sections of rivers in Japan.

PLAN FORM AND PROFILE FORM

Since plan form and profile form are both properties of hillslope form, it may seem inappropriate to discuss their relationship within the context of local controls of hillslope form. However, in the majority of hillslope studies the two properties have been considered in isolation (the latter to a much greater extent than the former) such that it is pertinent to examine the extent to which one acts as a local control of the other.

There has been no quantitative analysis of local variation in hillslope plan form. Along streams that have no appreciable valley floors, plan convexity on one side of the stream is matched by plan concavity on the other so that the two proportions must remain equal. Where a stream possesses a substantial floodplain, however, no such equality is necessary. Local weaknesses and concentration of joints within the underlying rock might be expected to lead to more rapid local backwearing of the hillslope, yielding a local concavity. The significance of this process for local hillslope plan form remains a matter of conjecture. Controls of this nature must, nevertheless, be regarded as the most likely determinants of local hillslope plan form and there is no reason to expect that it is affected by profile form. Traditionally, the converse has not been regarded as the case.

It has been widely assumed that plan curvature affects hillslope processes (e.g. Carson and Kirkby, 1972, pp.390-97; Young, 1972, p.177), and thereby affects hillslope profile form. This relationship is implicit in many models of hillslope profile development (e.g. Kirkby, 1986). The field evidence for the relationship is much less unequivocal. In a study of hillslopes in the dip slope of the South Downs, Parsons (1979) found no significant relationship between hillslope plan curvature and two measures of profile curvature. He concluded that if plan curvature does affect hillslope processes and is thereby related to profile curvature, it must do so at a very localised level and that this effect may not be so important as to affect overall profile curvature. The importance for hillslope processes of plan curvature is most closely defined for wash processes where there is a more than linear dependence on drainage area. Parsons's study may

101

be indicative of a weak relationship between plan form and profile form in a locality where wash processes are less important than soil creep. The same finding may not apply in environments where wash processes are dominant.

HILLSLOPES IN DRAINAGE BASINS

Within drainage basins all of the controls discussed above operate to yield an astonishing variability in hillslope form. It remains a challenge to geomorphologists to sort out the importance of these controls and to explain the variability. In a study of the variability in hillslopes within seven first-order drainage basins, Parsons (1982) examined twelve attributes of profile form. For profiles spaced at 200 m intervals he found coefficients of variation as high as 167% and calculated, on the assumption of spatial randomness in profile form, that spacing profiles as close as 3 m may be necessary to obtain reliable estimates of the mean values of some of these attributes (Table 6.1).

Attempts to explain variation in hillslope form within drainage basins have focused on changes in hillslope gradient (either maximum or mean), particularly on changes in this property in the downstream direction. Strahler (1950) argued that for equilibrium to exist between valley sides and stream channels the gradients of the two should be positively related. Thus steeper valley sides should contribute more sediment to the channels at their bases which, in turn, require steeper gradients to be able to transport it. Using data taken from widely spaced sites across the United States, Strahler was able to support this argument and obtained a relation between valley side gradient S_g and stream gradient S_c of

$$logS_g = 0.60 + 0.80logS_c$$

Although this argument seems to be valid when comparing one valley side with another, factors of relief and basal conditions seem to be more important than local channel gradient when attempting to explain the variation in hillslope gradient along individual valley sides. A number of authors (Arnett, 1971; Carter and Chorley, 1961; Summerfield, 1976; Frostick and Reid, 1982) have identified increasing stream gradient with increasing stream order. Both Carter and Chorley

102

Table 6.1: Distances (in metres) between adjacent hillslope profiles in a systematic sample necessary to determine mean values (to within 10% of the true mean with 90% confidence) of hillslope attributes in seven drainage basins (assuming spatial randomness in attributes of hillslope profiles)

	Drainage basin						
Hillslope attribute	1	2	3	4	5	6	7
Total length (S1)*	27	75	68	114	63	40	50
Vertical height (S2)	28	96	90	131	63	28	19
Curvature (Sh1)	63	56	129	136	65	12	12
Distance of maximum gradient from the crest as a percentage of S1 (Sh2)	400	343	277	378	367	76	67
Mean gradient (G1)	97	800	172	200	245	58	36
Maximum gradient (G2)	50	142	82	262	220	74	22
Percent profile length $>-2^o - <2^o$ (G3)	10	13	15	27	16	6	3
Percent profile length $2^o - <5^o$ (G4)	33	42	45	51	57	21	30
Percent profile length $5^o - <10^o$ (G5)	52	219	63	55	57	21	25
Percent profile length $10^o - <18^o$ (G6)	8	4	14	18	19	3	6
No. of changes of direction of curvature (R1)	14	12	20	37	5	13	12
Mean difference of gradient between adjacent segments (R2)	24	142	103	142	220	42	27

*Denotes the equivalent attribute in Table 3.1

After Parsons (1982, p.75), Earth Surface Processes and Landforms, 7. (*Reproduced by permission of John Wiley & Sons Ltd. Copyright 1982*)

and Frostick and Reid considered that their data came from stream networks that were actively expanding and argued that the increases in gradient that they observed as they sampled streams of higher order could be interpreted as temporal sequences. Richards (1977) considered such an interpretation to be invalid for Arnett's and Summerfield's studies. In a larger stable basin there would be insufficient evidence for development of the stream network, such that a higher order stream could not be assumed to have progressively increased its order through time. In consequence, the changes in hillslope gradient associated with increasing stream order must be explained by consideration of equilibrium relationships between hillslope and stream processes. Nonetheless, Richards argued, both actively expanding small stream networks and larger, more stable ones are subject to the same geometric constraints. In both cases increase in stream order is associated with greater depth of incision as the discharge of the basal stream increases. Thereby, slope length and relative relief both increase but the latter does so more rapidly than the former because drainage density is likely to remain more or less uniform within a catchment so that variation in hillslope length is more constrained. Geometrical considerations therefore require an increase in mean gradient. Carter and Chorley observed that gradient cannot continue indefinitely to increase with order. Once the hillslope reaches the angle of repose of the underlying material further increase in gradient is prevented. In fact, it has been generally found that hillslope gradient reaches a maximum in third- to fifth-order drainage basins but declines thereafter (Figure 6.5). This reversal of the

Figure 6.5: Relationships of mean valley-side gradient to stream order

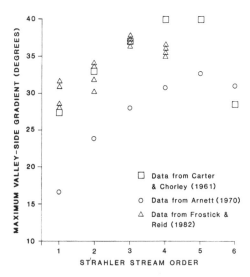

relationship has been attributed to the declining influence of the basal stream on hillslope gradient in higher order basins. Streams of higher order typically show evidence of lateral

104

shifting and a developing floodplain. Thus it has been argued that their influence on hillslopes is intermittent, at most, allowing time for hillslope processes to effect a reduction in hillslope gradient.

Hirano (1972a) found no relationship between hillslope gradient and stream gradient along individual valley sides but, like the previous authors, identified an increase in gradient as the amount of lowering of the stream bed increased. However, he assumed the steepness of valley sides to be proportional to the *rate* of lowering of the stream bed rather than to the *amount* of lowering. Although Hirano offered no evidence to support this assumption, it presents an interesting challenge to the more widely held view.

Changes of hillslope gradient along a stream of given order have been investigated by Parsons (1987). Like the studies that have examined changes in valley-side gradient with changing basin order, Parsons identified an increase in hillslope gradient with increasing distance from the stream head. To some extent, the geometric controls that Richards showed to apply as stream order increases also appear to operate along an individual valley side. Provided channel gradient is greater than the gradient along the drainage divide, there will be a downvalley increase in relative relief. However, Parsons also identified downvalley increases in gradient along valley sides where no such change in relative relief was present. Furthermore, in almost all cases, Parsons observed that the maximum hillslope gradients were not adjacent to the basal stream but separated from it by a low-angle footslope concavity. He argued that, in the absence of increasing relief in the downstream direction, the only available explanation for increased gradient is increased lateral cutting by the basal stream. Although this process is intermittent (as evidenced by the footslope concavities) it seems to be sufficiently important to affect hillslope gradient. This explanation appears to stand in contrast to the claims that lateral floodplain development accounts for the decrease in valley-side gradient in higher order basins (Carter and Chorley, 1961; Arnett, 1971; Frostick and Reid, 1982). On the other hand, it is consistent with theoretical studies of the effect of lateral undercutting of hillslopes (Scheidegger, 1960). The two explanations may, however, be reconcilable. If, in drainage basins of low order, undercutting, though intermittent, is frequent and effective it may be the dominant control of hillslope gradient. If, in contrast, the process is less frequent and less effective in higher order drainage basins, hillslope processes may be the determinant of gradient. At the present time there is no evidence concerning rates of lateral shifting of streams of different orders so that such a reconciliation remains no more than speculation.

Although the principal concern of research into variation of hillslope form within drainage basins has been the study of hillslope gradient, profile curvature has also been examined. Arnett (1971) found increasing profile convexity up to stream order four and a decline through orders five and six. Mathier and Roy (1984) identified an abrupt change in hillslope convexity along the valley side of a miniature stream. In both studies the changes in convexity were attributed to the effects of downcutting by the basal stream. Increased downcutting produces increased hillslope convexity.

An important aspect of the study of hillslopes in drainage basins is the question of valley asymmetry. Although not exclusively due to differences in microclimate across valleys, this has been the most widely used explanation. Accordingly, discussion of valley asymmetry is given within chapter 4.

FORM & PROCESS 7

For hillslopes, the relationship of form to processes depends on the extent to which processes are able to fashion hillslopes in accordance with their capacities for sediment transportation. Gilbert (1877) made a distinction between those situations in which the rate of weathering limits surface degradation and those in which it is the conditions limiting transportation that determine the rate of degradation of the land surface. More recently, this distinction has been used to dichotomise hillslopes into those on which processes are *weathering-limited* and those on which they are *transport-limited* (Carson and Kirkby, 1972, pp.104-6; Selby, 1982a, p.199). Such a dichotomy is unwarranted for it both implies that so long as weathering processes yield an ample supply of transportable particles transportational processes will always operate at capacity, and denies the possibility of active erosion by hillslope processes.

An alternative perspective is provided by Ellison (1947) who, in discussing transportation by hydraulic processes, identified *detachment* as a control of the rate of sediment removal. From the point of view of understanding the relationship of form and process on hillslopes, it may be more useful to regard all surface degradation as subject to detachment control. The surface materials of hillslopes may be thought to lie along a continuum of detachability with respect to a particular hillslope process (Figure 7.1). At one end of the continuum surface materials are infinitely detachable so that surface degradation is limited by the transportation capacity of that process. At the other end of the continuum surface materials have zero detachability and

107

removal is limited by the rate of weathering. Between these two endpoints, detachability is affected by such factors as soil aggregate stability, surface crusting and pore pressure.

Figure 7.1: The detachability continuum

With respect to a particular process acting on a hillslope, detachability varies both in space and with time. Differences in soil composition, for example, may exist down a hillslope and so alter the ease with which that soil can be entrained. Processes that operate sporadically may have a supply of readily transportable material on one occasion and hence operate at the transport-limited end of the continuum, but at a different time or season the surface material may be less easily entrained. Equally, a particular hillslope acted upon simultaneously by many processes may present surface materials that appear at a different point on the continuum to each process.

Inasmuch as processes fashion hillslopes, it might be assumed that the relationship of form to processes is no more than one in which the latter controls the former. This assumption underlies process-response modelling of hillslope form. Such a view is an over-simplification and is one that fails to take account of temporal control in cause-effect relationships (Schumm and Lichty, 1965). Because of feedback mechanisms in the long-term control of process on form, in the short term it is form that determines the effectiveness of process. Understanding both control mechanisms is important. Long-term control explains the manner of hillslope evolution and hence helps to account for the hillslope forms present on the earth's surface: short-term control explains how hillslopes respond to particular process events and can be useful in planning man's utilisation of natural hillslopes and in design of artificial ones.

In this chapter it is useful to adopt the methods of process-response modelling to examine the nature of process-form relationships. The basis of this approach is the continuity equation, which in its simplest form is

$$\frac{\delta y}{\delta t} = \frac{\delta S}{\delta x} \qquad (7.1)$$

This equation states simply that for any point on a hillslope the change in elevation (δy) with the passage of time (δt) equals the change in sediment transport (δS) as distance along the hillslope changes (δx). If sediment transport at any point can be expressed as a function of the gradient of that point (denoted by the tangent of the angle, which is given by $\delta y/\delta x$) and the length (x) of the hillslope upslope of that point so that

$$S = f(x).(\delta y/\delta x)^n \qquad (7.2)$$

and if it can be further assumed that sediment delivered to the base of the hillslope is subject to unimpeded basal removal, then an approximate solution to equation 7.1 can be obtained wherein

$$y_0 - y \propto \frac{\{x\}^{1/n}}{f(x)}.\delta x \qquad (7.3)$$

in which y_0 is the elevation of the hillslope divide. This approximation defines what Kirkby (1971) termed the *characteristic form* for a particular process because the form of the hillslope given by this equation is the one to which any initial form will asymptotically tend under the operation of that process.

In the following section the hillslope forms that may be expected from the operation of particular processes are examined. In reality, no hillslope is subject exclusively to the operation of one process and more than one process may lead to the same hillslope form. Nonetheless, it is instructive to consider separately the implications of each process for hillslope form.

HILLSLOPE PROCESSES AND PROCESS-FORM RELATIONS

Hillslope processes are conventionally divided into those dominated by water flow forces, termed *wash processes*, and those dominated by gravitational force, termed *mass movement processes*. Each class of processes encompasses great variability and the distinction between the two classes is largely a matter of convenience. Carson and Kirkby (1972, p.99) used the relationship of neighbouring particles as an indicator of the type of movement. The characteristic of mass movement is that neighbouring particles remain close together as they move so that it is appropriate to consider movement in terms of forces acting on the mass as a whole. By contrast, in water flow particles move as individuals with little or no relation among neighbours, so that movement can be understood in terms of the forces acting upon individual particles. Nevertheless, in some processes conventionally classed as mass movements the water component of moving debris may be so great that the relations among neighbouring particles is almost as weak as in wash processes.

Wash Processes

Rainsplash, overland flow, subsurface flow, and solution are the hillslope processes that depend upon water for their denudational activity and hence are classed together under the somewhat unsatisfactory term *wash processes*. In reality, it is difficult to separate the effects of the four processes because a storm event will cause some, if not all, of them to act simultaneously on a hillslope. Their mechanisms are, however, very different and it is useful to evaluate separately the effect of each process on hillslope form. Although any source of water can contribute to solution loss, for the purposes of this analysis the solution process will be considered separately. The discussion of rainsplash, overland flow and subsurface flow will consider only their ability to remove undissolved material from hillslopes.

Rainsplash
The energy of falling raindrops can effect movement of particles on hillslopes in two ways directly, and in a further way indirectly. Raindrops falling onto a bare rock surface will break up on impact and the smaller droplets will rebound around the initial site of impact. If there are small loose fragments on the rock surface some of these may be entrained in the rebounding droplets and thus be transported away from the impact site. On an inclined surface this process will lead to a net downslope transportation of the loose particles because the

110

droplet trajectories will be longer on the downslope side. On a soil surface the same process will operate but, additionally, the raindrops may have sufficient energy to crater the soil surface thus pushing soil particles outward from the impact site. This process will yield an asymmetrical crater if the direction of impact is other than normal to the surface. For vertical or near vertical raindrops the asymmetry is most likely to favour downslope movement. By moving small particles in these ways, raindrops may move larger ones indirectly by undermining them.

For a raindrop of given size and impact velocity, the ability to effect net downslope movement of surface particles depends on the supply of suitably sized particles and on hillslope gradient. Assuming that there is an abundant supply of particles (i.e. the process is transport-limited) then the sediment transport rate, S is given as

$$S = (\delta y / \delta x)^n$$

and equation 7.1 reduces to

$$y_0 - y \propto x^{2/n}$$

Theoretically, therefore, hillslopes dominated by rainsplash will develop convex profiles. In reality, of course, it is very unlikely that rainsplash alone could fashion a hillslope. The rainfall would give rise to other wash processes which are likely to have greater effect on the form of the hillslope. Nevertheless, close to hillslope divides rainsplash may be the dominant process. Although theoretical and experimental work support the view that hilltop convexities could be formed by rainsplash (Mosley, 1973), other mechanisms have been proposed (e.g. Gilbert, 1909; Yair, 1973). So far, no study has been undertaken to measure the relative importance of all processes acting on hilltop convexities to establish which, if any, is dominant.

Overland Flow
Overland flow can be divided into two types: infiltration-excess flow (sometimes termed *Horton* overland flow) and saturation flow. Horton (1945) developed a model of overland flow on soil-covered hillslopes based on the assumption that rainfall intensity exceeds the infiltration capacity of the soil, which he took to be more or less uniform over the entire hillslope. Horton argued that

initially the excess rainfall would fill small depressions in the irregular microtopography of the soil surface but that these depressions would soon be overtopped and that water would begin to flow down the hillslope. Although Horton recognised that much of the flow would be concentrated into rills, the distinction between rill flow and interrill flow was not incorporated into his determination of the eroding force of overland flow, which was calculated on the assumption of a water layer of uniform thickness across the hillslope so that

$$d_x = x. \frac{r}{v}$$

where d_x is depth of flow, x is the distance from the divide, r is runoff rate (rainfall - infiltration) and v is velocity. Furthermore, since the force of erosion of the water at distance x, F_x, can be determined from

$$F_x = \gamma_w.d_x.S$$

where γ_w is the unit weight of water and S is the gradient, a general expression can be derived linking the force of erosion to distance from the divide and gradient:

$$F \propto f(x).S^n,$$

where n is an exponent.

The effect of Horton overland flow on hillslope form can thus be examined within the framework of equation 7.2 and a characteristic form (equation 7.3) can be identified. Empirical studies of erosion from runoff plots and cultivated fields have indicated values of about 1.35 for n, and for $f(x)$ to be given by x^m with the exponent m being about 1.5 (Zingg, 1940; Musgrave, 1947; Kirkby, 1969). The characteristic form is concave throughout.

There are reasons to question the relevance of these empirical inputs and hence the validity of the characteristic form. Mention has already been made of the fact that on natural hillslopes much overland flow is concentrated into rills whereas Horton's analysis assumes a sheet of water of uniform thickness across the hillslope. It has been commonly assumed that rills represent ephemeral concentrations of erosive activity on hillslopes and that these concentrations move around so that the whole hillslope is lowered at a more or less uniform rate (e.g. Carson and Kirkby,

1972, p.226). Dunne and Aubrey (1986) have argued that overland flow alone is unstable and produces a rilled surface and that during an individual rainstorm it is the associated rainsplash which counters or reverses the tendency to form an incised and integrated channel network. In between rill-forming rainstorms, rainfall of lower intensity and other processes may transfer sediment into the rills so that, even if rills do remain in fixed positions over long periods of time, sediment moved down them may be derived from the entire hillslope surface. The empirical inputs to the process-response model for overland flow have been obtained either from individual rainstorm events, in which the effects of rainsplash are included, or from longer periods of observation, in which inter-storm processes will have contributed to erosion. In either case the results cannot be used to model the effects of Horton overland flow acting alone.

Horton's assumption that infiltration is spatially uniform is unlikely to hold. Downslope, soils are likely to become finer-textured and hence have lower infiltration capacities. The effect of this, however, would be simply to make the downslope increase in depth of overland flow more rapid than otherwise. This would increase the exponent m but the characteristic form would remain concave. The more serious problem lies with other spatial irregularities in infiltration. There is evidence to show that infiltration on hillslopes exhibits considerable spatial variation. The effects of this irregularity are difficult to model but if the irregularities are large they may have sufficient impact to override any simple dependency of F on distance and gradient.

The second model of overland flow (described by Kirkby and Chorley, 1967 as the *saturation overland flow model*) identifies conditions under which overland flow occurs even though rainfall intensity is less than the infiltration capacity of the soil. Once all pore spaces within the soil layer are water-filled and no more water can enter, all rainfall becomes infiltration excess and contributes to overland flow. The length of time needed for this to occur will depend on the total volume of soil and its water content prior to the onset of rain. In consequence, saturation overland flow will occur most readily in areas of thin soils or where initial water content is high.

The former are likely to occur on upslope sites or on plan convexities. Flow off such areas may infiltrate downslope or may add to existing Horton flow. Low-angle, footslope areas adjacent to streams are the most likely areas of high antecedent soil moisture. The small gradient and short distance of travel to the foot of the hillslope suggest that saturation overland flow generated from these areas will have low capacity for sediment transport.

Spatial heterogeneity is the essence of saturation overland flow. The process is so strongly dependent upon the particular character- istics of individual hillslopes that the idea of a characteristic form for this process is probably meaningless. Once saturation overland flow is generated it will be subject to the same laws of hydraulics as Horton overland flow but it cannot always be assumed that depth of flow increases downslope.

Subsurface Flow
Water that infiltrates into the soil is subject to the same determinants of its continued downward passage as that arriving at the soil surface. It will continue downward so long as the supply rate is no greater than the downward infiltration capacity of the soil. Where this is not the case, either because the lower soil layers are saturated or have a lower infiltration capacity, there will be an infiltration excess. A sudden reduction in infiltration capacity is common at soil horizon boundaries, particularly at the boundary between upper eluviated horizons and lower illuviated ones. Because soil horizons tend to be more or less parallel to the ground surface there will be a tendency for excess water to move downslope along horizon boundaries. Two types of subsurface flow may be identified, termed *throughflow* and *pipeflow*. Throughflow consists of water percolating through pores within the soil. Velocities of throughflow are low (a few metres/hour, at most) so that the capacity for sediment transport is limited. It has been widely believed that throughflow carries very little sediment but work by Pilgrim, Huff and Steele (1978) contradicts this view. These workers observed suspended sediment concentrations in throughflow of over 1000 mg/l which they noted to be as high as many surface streams during moderate storm-runoff events. Because sediment is not

114

transported through pores similar in size to the transported particles, Pilgrim *et al.* concluded that the water must be travelling through macropores in the soil. Nevertheless, the low velocities of such flow preclude detachment and entrainment by such flow and the authors inferred that the sediment in such flow derives from the soil surface, having been detached by raindrop impact.

The effect of throughflow on hillslope form is difficult to model. Accepting the inferences of Pilgrim *et al.* regarding the origin of sediment transported by throughflow implies that sediment removal by this process may not be transport limited, particularly towards the lower parts of hillslopes. However, if this problem is disregarded, the probability for any storm event is that subsurface discharge increases downslope and a characteristic form similar to that for surface flow, i.e. concave throughout, results.

In pipeflow the velocity of subsurface water is much greater: values as high as 1 metre/sec have been recorded. Pipes show a great range in size, from no more than a few mm in diameter (where they differ very little from the macropores of diffuse throughflow) to as much as a few metres. The role of pipeflow in channelling runoff from hillslopes has been recognised only relatively recently and there is still a paucity of data. Nonetheless, it is becoming evident that pipeflow is extremely widespread and there are indications that in some environments it may be the dominant form of water movement on hillslopes (Jones, 1982). Unlike throughflow, pipeflow is clearly capable of entraining and transporting sediment off hillslopes and Bryan and Yair (1982) have asserted that in some badlands pipeflow may also be the dominant denudational process.

The role of pipeflow in fashioning hillslopes can, at present, be no more than guessed at. The major difficulty lies in understanding how concentrated linear flow affects entire hillslope areas (Jones, 1987). In some badlands there is evidence of former large pipes having collapsed to create surface gullies and it might be imagined that pipeflow has transformed an initial plan planar hillslope into a dissected one. A more attractive view may be the model presented by Haigh and Rydout (1987). This model proposes a sequence of pipe and gully formation illustrated in Figure 7.2. Haigh and Rydout's model suggests reformation of the pipe in the same location but this may not be the case. In any event some areal lowering may be achieved.

Figure 7.2: A model for pipe and gully formation. (*After Haigh & Rydout, 1987,* International Geomorphology 1986, *V. Gardiner ed. Reproduced by permission of John Wiley & Sons, Ltd. Copyright 1987*)

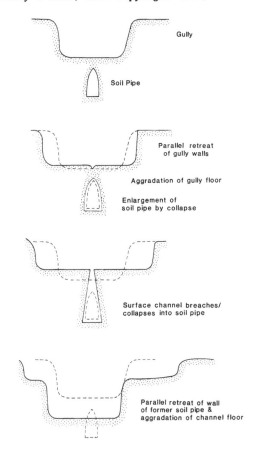

Gully

Soil Pipe

Parallel retreat of gully walls

Aggradation of gully floor

Enlargement of soil pipe by collapse

Surface channel breaches/ collapses into soil pipe

Parallel retreat of wall of former soil pipe & aggradation of channel floor

Solution

All water that comes into contact with hillslope materials may contribute to loss of minerals in solution. A critical determinant of the amount of material removed is the length of time water is in contact with the minerals compared to the length of time necessary for equilibrium to be

116

approached. If the former is greater than the latter the water will dissolve the maximum possible amount (in this case the process may be regarded as transport-limited). Kirkby (1978) has argued that most soluble rock material approaches equilibrium with water in days rather than minutes and that, in consequence, for most minerals it is only slow-moving subsurface flow that leads to appreciable removal in solution. He has used this argument to propose that removal in solution can be considered wholly in terms of subsurface throughflow. Thus in humid areas, where overland flow increases downslope and a decreasing proportion of rainfall adds to subsurface flow, solution removal lessens in the downslope direction.

A difficulty in extrapolating from rates of solution to characteristic hillslope forms lies in equating loss of material with denudation. The differences in specific gravity between fresh and weathered rocks imply that substantial amounts of solutional loss may result in little volumetric change. Thus in rocks that contain a substantial insoluble residue, much solutional loss may be achieved without significant landform change. Likewise, deposition of solutes by precipitation may fill existing voids rather than cause aggradation.

The exception is limestone. Most of the constituents of this rock type are soluble, hence solution and landform change can be equated. In addition, reaction times are much shorter, so that all water that comes into contact with limestone may achieve some dissolution. Properties of the soil cover, however, are of considerable importance. Trudgill (1985, p.123) argued that a common situation is one of more acid soils on the upper parts of hillslopes so that dissolution is concentrated here and hillslopes are subject to the sort of hinge decline proposed by Carson and Kirkby (1972, p.260). Variations in soil thickness accompanying variations in acidity may modify this pattern.

Mass Movement Processes

There have been a number of attempts to classify the range of processes that can be identified under the heading of mass movements (Sharpe, 1938; Varnes, 1958; Hutchinson, 1968; Carson and Kirkby,

1972). These attempts have classified mass movement processes in terms of the type and rate of movement and the type and water content of moving material. Even in these terms, none of the classifications has been wholly successful since none has considered all four of these variables independently. One of the most successful is Carson and Kirkby's (Figure 7.3) which identifies mass movement processes as a function of three of

Figure 7.3: Classification of mass movement processes. (*After Carson & Kirkby, 1972,* Hillslope Form and Process, Cambridge University Press. *Reproduced by permission*)

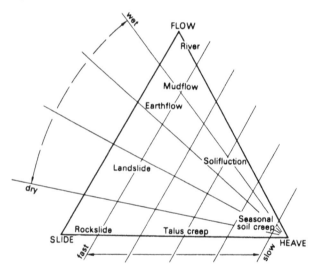

the variables. The strength of this class-ification is that it recognises that any mass movement process may result from a type of movement that has some of the qualities of three types of movement included in the classification: slide, flow and heave (Figure 7.4). The weakness of the classification is that it is incomplete because it fails to take account of the type of material as an independent variable and it identifies only three types of movement. Other classifications recognise *fall* as another type of movement. Furthermore, heave is not a type of downslope movement of material as are flow, slide and fall. It involves displacement upward and normal to the slope. Heave facilitates downslope displacement, but itself does not directly cause it.

Figure 7.4: Velocity profiles for ideal mass movement types. (a) Pure slide. (b) Pure flow. (c) Pure heave. (*After Carson & Kirkby, 1972, Hillslope Form and Process, Cambridge University Press. Reproduced by permission*)

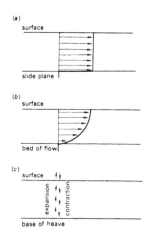

Mass movement processes dominated by the heave mechanism tend to be slow-acting and are best considered separately. The other mechanisms effect relatively rapid downslope transportation of material. The processes dominated by these mechanisms can all be considered with reference to the relationship between the force G tending to cause material to move and the force R acting to resist that movement. G can be accounted for by the equation

$$G = m.g.\sin\alpha,$$

where m is the mass of the potentially moving material, g is the force of gravity and α is the gradient of the plane on which the potentially moving material is resting. R can be accounted for by the equation

$$R = c + \sigma' \tan\varphi,$$

where c is the cohesive force, σ' is the effective pore pressure, and φ is the angle of internal friction. The ratio R:G determines the *factor of*

119

safety of the potentially moving material. For values of this ratio $\geq$ 1 the mass of material is stable and movement should not occur. For values < 1 the mass is unstable. The nature of the resulting mass movement process may, in part, depend on the way in which the inequality

$$m.g.\sin\alpha > c + \sigma'\tan\varphi$$

has been brought about. Factors which bring about an increase in the value of the left-hand side of the inequality but not an accompanying decrease in the right-hand side, tend to promote processes dominated by the slide mechanism. Factors that decrease the value of the right-hand side are likely to result in processes dominated by the flow and fall mechanisms.

Figure 7.5: Mass movement source area and deposit. In this earthflow the concave source area and the convex deposit partly overlap

Whichever mass movement process operates, however, the result is the same. The moving mass of material continues downslope until it is either removed from the hillslope entirely or it comes to rest at some point where $F \geqslant 1$. In the latter, more common case a mass of material will, relatively rapidly, have been removed from an upslope area and deposited in a downslope one. Because the mass movement process operates on a finite mass of material, the upslope source area of this material is typically concave in plan and profile whereas its downslope resting place is usually convex in plan and profile (Figure 7.5). Depending on the mass movement process and the translational and rotational components of the movement, the concavity and convexity may be well separated or they may overlap. Where the two areas are well separated the path over which the material has moved will have features that vary with the mass movement process (Figure 7.6). Hillslopes dominated by mass movement processes are characterised by numerous such concavities and convexities and hence display great irregularity of form in plan and profile (Figure 7.7).

Figure 7.6: Features of a debris-flow channel. The path between the well separated source area and deposit of this debris-flow shows characteristic levées and a hummocky channel floor

Figure 7.7: Hillslope form dominated by the effects of mass movement processes. The results of several individual mass movements can be identified within the overall plan and profile irregularity of the hillslope

The inequality $m.g.\sin\alpha > c + \sigma'\tan\varphi$ describes the conditions necessary for mass movement but, because it applies to a finite mass of material, it is less straightforward to determine the hillslope form that results from the operation of mass movement processes. This determination is easiest if there are infinitely many mass movements and each involves only a small quantity of material. In this case the process can be modelled as a continuous rather than a sporadic one, and as one that affects the entire hillslope rather than only parts of it. This type of model can be used to predict the development of a scree slope under rockfall. Statham (1976) obtained good agreement between the characteristics of scree slopes on the Isle of Skye and those predicted from a model based on rockfall from a headwall associated with transport of particles over the scree slope. This agreement included not only the gradient properties of the scree slopes but also variation in gradient with the height of the headwall and characteristics of particle size distribution.

Iida and Okunishi (1983) considered the long-term development of hillslopes on which landsliding is the dominant process. They argued that although landslides are sporadic and local events they represent no more than the more obvious and dramatic elements of a continuous process. They contended that a landslide will occur at a particular locality when the thickness of the weathered layer attains a critical thickness, which itself is a function of gradient. Although this approach may be useful in predicting the occurrence of a landslide at a particular locality, it cannot be used to relate hillslope form to landsliding. The mass movements occur as a function of weathering rate and gradient, neither of which are controlled by the mass movement process. The model simply describes the process mechanisms operating on steep hillslopes where the rate of weathering is an important determinant of the effectiveness of processes.

Previous workers have also emphasised the role of weathering in controlling the form of hillslopes dominated by mass movement processes. Carson (1969) proposed that weathering leads to continuous changes in the character of the soil mantle and that these changes are associated with phases of mass movements. During each phase, mass movement processes cause significant reductions in hillslope gradient. The mantle remains temporarily stable at these new, lower gradients until such time as weathering processes alter the character of the mantle and a new phase of mass movements is initiated. Each reduction in gradient by mass movement processes yields a new threshold of stability in the soil mantle that weathering processes must exceed before the next phase of mass movements can be initiated. The resulting hillslopes are ones in which these threshold gradients commonly form significant rectilinear segments. If a rectilinear segment on a hillslope is inclined at the threshold of stability for its soil mantle then its gradient α can be determined from

$$\alpha = \arctan[(1 - m.\gamma_w/\gamma_s)\tan\varphi_r]$$

where γ_w is the unit weight of water, γ_s is the saturated unit weight of soil, φ_r is the residual angle of shearing resistance, and m is the position of the water table with respect to the ground surface (taken to equal 1, i.e. at the ground surface). This determination assumes that cohesion is zero (taken to be the case for long-

123

term stability). Carson and Petley (1970), in the southern Pennines and on Exmoor, and Rouse and Farhan (1976), in South Wales, have applied this concept to rectilinear segments on hillslopes and concluded that the gradients of these segments are consistent with their being the limiting angles of stability for the soil mantles.

Not all parts of a hillslope are equally likely to be sites of mass movements. Inasmuch as a temporarily high water content often triggers a mass movement process, it might be anticipated that plan concavities would be preferential sites for mass movements. Empirical data appear to support this notion (Iida and Okunishi, 1983). Although this observation indicates that the likelihood that a mass movement will occur at any location on a hillslope is strongly influenced by the plan curvature of that location, it does not demonstrate that hillslopes dominated by mass movement processes will develop pronounced plan curvature. Dietrich and Dorn (1984) have shown how landslides evacuate hillslope hollows that have been filled by colluvial processes. Empirical data showing concentrations of mass movements in hollows may be no more than observations of this type of process. Nevertheless, using modelling of weathering processes and subsequent mass movements, Iida and Okunishi have argued that the rate of denudation is greater in plan concavities than on convexities so that mass movement processes accentuate any initial hillslope plan curvature rather than eliminate it.

The Heave Mechanism
Heave is the displacement of hillslope material upward and normal to hillslope gradient. The main causes of displacement (in order of importance in a humid temperate climate, according to Kirkby, 1967) are:

  expansion due to wetting
  expansion due to freezing
  displacement by fauna (principally earthworms and burrowing animals)
  thermal expansion
  displacement by plant roots.

Heave is followed by settling as soils dry, ice melts, and so on. The heave mechanism results in downslope movement because of the disparity in the directions of heave and settling. For an isolated particle, it is argued that heave is normal to the surface whereas settling is vertical; hence in one

124

cycle of heave and settling the particle will move downslope by the distance of heave times the tangent of hillslope gradient (Figure 7.8A). However, with few exceptions, particles are not free to move simply in response to these mechanisms but are constrained by surrounding particles. Indeed, much experimental work suggests that the heave mechanism operates on soil masses rather than on individual particles, which is the distinguishing characteristic of mass movements (Carson and Kirkby, 1972, p.99). In consequence, the disparity between the directions of heave and settling is rather less so that net downslope movement is less (Figure 7.8B).

Figure 7.8: Trajectories for particles subject to heave
  A. Isolated particle is lifted normal to the hillslope surface and falls vertically so that downslope movement is $d.\tan\alpha$
  B. Particle within a mass is lifted normal to the surface but is prevented from falling vertically by neighbouring particles. Hence it returns to the hillslope closer to its original position and downslope movement $d.\tan\alpha$

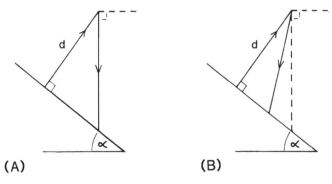

(A)   (B)

The amount of downslope movement achieved in any heave-settle cycle is extremely small. Hence the effects of this mechanism on hillslope form can be modelled as a continuous process. Like rainsplash, the effectiveness of the process is determined by gradient so that the characteristic form for this process can be given by

$$y - y_0 \propto x^{2/n} \tag{7.4}$$

where n is a function of the extent to which

125

interlocking of particles reduces the direction of settling below 90°. This characteristic form assumes that the potential for the heave mechanism is equal throughout the length of a hillslope. Young (1972, p.51) has pointed out that this potential is dependent on properties of the regolith which may be expected to vary systematically. In general, regolith becomes finer-textured and wetter downslope, hence the potential for the heave mechanism may increase downslope.

The heave mechanism is widely taken to be synonymous with the process of soil creep. An alternative view is taken by Culling (1963) who has modelled soil creep as a diffusion process acting at the level of individual soil particles. Culling argued that the effect of temperature changes, soil fauna and the like is to impart random velocities to soil particles. Because of the close proximity of soil particles they are seldom able to move in proportion to the force applied; rather their movement is governed by the availability of pore spaces into which they may move. In addition to these acquired random velocities, non-random effects of gravity and water seepage operate, so that there is a directional bias to the movements of soil particles. Taking this stochastic approach to the analysis of the movement of individual soil particles, Culling identified a characteristic form for hillslopes under the action of soil creep similar to that given in equation 7.4.

FORM-PROCESS RELATIONS

The conclusion reached by Arnett (1971, p.81) that "slope form appears to determine contemporary geomorphic process" indicates that the relationship of process to form is more than simply one of cause to effect. Equations of the form of 7.2 define links between geomorphic processes and hillslope form that operate irrespective of whether or not processes are the principal determinant of hillslope form. Whilst such equations can be manipulated to yield the characteristic hillslope form that will develop in the long term under the operation of a particular process (equation 7.3), they demonstrate, above all, that the operation of processes depends upon existing hillslope form. This is important for two reasons. First, recent earth history suggests that even if it can be shown that present-day processes are transport-limited, sufficient time may not have elapsed for the characteristic forms to have developed. Hillslopes as they appear today may owe their form as much to processes that operated in the past as to those operating at the present time. The preservation of some of their history in the forms of hillslopes introduces difficulties into attempts to obtain empirical support for hillslope process-response models.

The second reason why the influence of form on process is important lies in its implications for the applications of hillslope geomorphology. In predicting, for example, sediment yield or the likelihood of landsliding, existing hillslope form is a crucial factor. Considerable attention has been given to this issue by workers in the U.S. Department of Agriculture, much of it in the context of predicting erosion by overland flow using the Universal Soil Loss Equation. Amongst the earliest work, Zingg (1940) proposed a relation between soil erosion E and hillslope gradient S and length L:

$$E \propto S^{1.4}L^{1.6}$$

Similar equations relating erosion rates to gradient and/or length have been developed by Smith and Wischmeier (1957) and Meyer and Monke (1965), among others. Bryan (1979) pointed to the fact that most studies of the effect of gradient on sediment entrainment have been confined to relatively gentle gradients. He studied variations in sediment entrainment on gradients ranging from 3° to 30° and concluded that the relationship between the two was better described by a polynomial, rather than a power, function. More recently attention has moved to consider not only hillslope gradient and length but curvature as well (Onstad, Larson, Wischmeier and Young, 1967; Young and Mutchler, 1967; Meyer and Kramer, 1968; Hadley and Toy, 1977). Meyer and Kramer applied equations for relating sediment removal to hillslope gradient and length (developed by Zingg and Smith and Wischmeier) in simulations of erosion on four hillslope profile shapes (rectilinear, concave, convex and convexo-concave). They found that minimum sediment loss occurred on the concave profile and that this profile shape was changed least by overland flow processes. The other three profiles were modified

127

in such a way as to introduce or increase concavity. The conclusions from this study were that where it is necessary to minimise erosion on natural hillslopes they may be refashioned to make them concave in profile and that artificial hillslopes (for example road embankments) should be created concave in profile. Hadley and Toy applied simulated rainfall to convex, concave and rectilinear sections of natural hillslopes and recorded maximum erosion from the rectilinear sections. They, therefore, argued that erosion control measures should be focused on these sections. Both of these studies considered only erosion by overland flow: other optimum profile forms can be expected to exist to minimise sediment loss by other hillslope processes. It may be noted that, in Mayer and Kramer's study, minimum erosion was predicted from a hillslope shape similar to the characteristic form for the process. This would suggest that hillslopes evolve under the operation of a particular process to a form that minimises the effectiveness of that process.

HILLSLOPES AFFECTED BY SEVERAL PROCESSES

Few, if any, naturally occurring hillslopes are fashioned solely by one process. The effects of several processes acting in concert need to be considered not only from the point of view of which process may be dominant at any point on a hillslope but also from that of interactions among the processes.

If the amount of sediment transport effected by each of the processes acting on a particular hillslope can be described by an equation of the form of 7.2, then the dominant process at any point on the hillslope can be readily identified. Discussion of individual process-form relations has already shown that not all processes are amenable to description by equations of this type. A second, and perhaps more important, difficulty lies in the fact that although several processes may act on a hillslope their effects are neither simultaneous nor independent. The former problem is discussed at greater length in the following section. The latter means that the equation developed for an individual process may need to be modified when that process acts in concert with others. Jahn (1968) has considered this problem within the context of what he terms surface and linear degradation. Surface degradation refers to those processes that effect uniform ground surface lowering (e.g. rainsplash), whereas processes of linear degradation cause selective lowering of the ground (e.g. gullying). If surface degradation affects the upper part of a hillslope and V-shaped gullies incise the lower part but only affect one quarter of the surface of the hillslope, then the rate of gully incision will need to be eight times that of surface degradation to maintain the balance of the hillslope. We can infer from Jahn's analysis that a consequence of the deep gully incision will be to increase greatly the surface area of the lower part of the hillslope. This increase will promote the effectiveness of processes of surface degradation. Dunne (1980) has pointed to the crucial interaction between rainsplash and overland flow in controlling the growth and decay of rills. Inadequate attention has so far been paid to feedback mechanisms of this sort among hillslope processes. This is particularly so with regard to the three-dimensional form of hillslopes. Spurs and hollows created by one process may have major effects on another, as discussed by Crabtree and Burt (1983), for example, for the case of solutional denudation.

THE PROBLEM OF MAGNITUDE AND FREQUENCY

The examination by Wolman and Miller (1960) of magnitude and frequency of forces in geomorphic processes led these authors to the conclusion that some landforms are controlled by events of moderate magnitude and frequency. They recognised, however, that their analysis was based almost exclusively upon depositional landforms and pointed out (p.72) that their study contained "a notable lack of examples demonstrating effect-iveness of moderate events of frequent occurrence in moulding erosional landforms". Notwithstanding Wolman and Miller's identification of this shortcoming in their work, other authors have found the notion that one size of event may dominate the landforming process so attractive as to seek to apply it more widely in geomorphology. For hillslopes Carson and Kirkby (1972, p.103) raised the possibility of specific associations between hillslope forms and processes.

Two major difficulties can be identified in attempting to apply Wolman and Miller's notion to hillslopes. The first is associated with the related concept of a dominant event. For rivers, Blench (1951) identified the *dominant* discharge as the "steady discharge that would produce the same result as the actual varying discharge". This discharge can be identified as the bankfull discharge. Since it is a matter of common observation that rivers with greater discharges have larger channels, to ask which value in the varying discharge of a particular river determines channel size is a relatively straightforward question. By contrast, there is no observational evidence to suggest, for example, that large raindrops lead, via rainsplash, to a characteristic hillslope form different from that produced by small raindrops, or that the effects of long duration storms are different from those of storms of short duration. Hence geomorphologists have ignored dominant raindrop size and storm duration in considering the effects of rainsplash on hillslope form. Instead the concept of a dominant event fashioning hillslopes has been applied quite differently. It has been used to examine which of several processes acting simultaneously is the dominant one. This is very different from Blench's idea of a dominant discharge because it is unlikely to be true that a process dominant in this sense and acting alone could result in the same hillslope form as the several processes acting together.

Young (1972, pp.85-8) raised the issue of the dominant process fashioning hillslopes by pointing out that a single landslide moving a volume of 10 m^3 a distance of 10 m achieves the same effect as a continuous process that has an average annual rate of 10 $cm^3/cm/yr$ operating for one million years. From this calculation Young argued that mass movements must dominate denudation wherever there are signs of their activity. However, he further observed that on many hillslopes denudation must be dominated by more continuous processes since they do not display any signs of the irregularity of form characteristic of mass movement activity. Other workers who have attempted to identify the dominant process on hillslopes have included Rapp (1960) and Aniya (1985). Rapp monitored all processes on mountain hillslopes in the Kärkevagge area, northern Scandinavia (Table 7.1) and concluded that removal

130

in solution was the dominant denudational process. By contrast, in the Amahata basin, Japan, Aniya concluded that nearly 50% of reservoir sedimentation rates could be accounted for by landslide volumes. It seems reasonable to expect the dominant process to be different in different environments. At the present time there is a paucity of data that might allow identification of dominant processes. Only the crude dichotomy into hillslopes dominated by sporadic mass movements and those dominated by more continuous process (creep and wash) proposed by Young, and based wholly upon morphological evidence, can be cited.

Table 7.1: Relative importance of denudational processes in Kärkevagge, 1952-60

Process	Removal tons/km²/yr	Vertical movement ton-metres
Rockfalls		
Pebble-falls	1	845
Small boulder-falls	1.7	4,160
Big boulder-falls	6	14,560
Avalanches		
Small avalanches	1.4	1,050
Big avalanches	14	20,800
Earth-slides etc.		
Bowl-slides	20	75
sheet-slides	23	20,000
Sheet-slides + mudflows	18	70,000
Other mudflows	8.4	6,300
Creep		
Talus-creep	--	†2,700
Solifluction	--	‡5,300
Running water		
Disolved salts	26	136,500
Slope wash	?	?

† Horizontal component of talus-creep = 4,700
‡ Horizontal component of solifluction = 19,800

Based on Rapp (1960, p.185), Geogr. Annal.,42A. *Reproduced by permission*

Where the concept of magnitude and frequency has been applied to a single process acting on hillslopes it has not been from the point of view of determining the dominant size of event so much

131

as to characterise the frequency of events of different sizes. The issue can be most easily examined for discrete events that can be readily identified from their morphological effects on hillslopes and/or from their deposits. Hsu (1983) presented a formal statement of the expected relationship of frequency and magnitude for landslides such that

$$f = k.V^{-1}$$

where f is the frequency, V is volume, and K is a constant. Empirical support for a relationship of this form is provided in the measurements of small rockfalls by Douglas (1980) over a 72-week period in Northern Ireland and by Gardner (1983) over a 7-year period in the Canadian Rocky Mountains (Figure 7.9A). Similar distributions were found for debris flows around Mount Shasta, California (Hupp, 1984), and in northwest Europe (Innes, 1985). However, the frequency distribution of larger rockslides in Iceland during the Postglacial (Whalley, Douglas and Jonsson, 1983) is very different (Figure 7.9B). Although there may still be some doubt as to the form of the

Figure 7.9: Frequency distributions of rockslides
  A. In Antrim, N. Ireland (*After Douglas, 1980, Earth Surface Processes and Landforms,5. Reproduced by permission of John Wiley & Sons, Ltd. Copyright 1980*)
  B. In Iceland (*After Whalley et al., 1983, Geog. Annal.,65A. Reproduced by permission*)

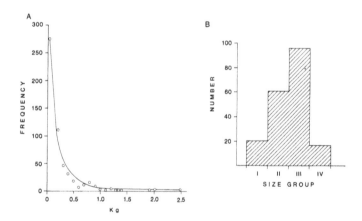

132

frequency distribution of mass movements, the evidence would seem to suggest that it is different from the log-normal frequency distribution given by Wolman and Miller (1960) for hydrological and meteorological events. If Hsu's model is valid, it implies that magnitude and frequency are more or less counterbalanced and no dominant size of event can be identified. These studies, however, consider quantities of material moved by different-sized events: they do not consider qualitative differences that may exist between smaller and larger mass movements. Nor do they consider the effect on hillslope form of events of different size.

This latter issue was addressed by Selby (1974a) who considered the effects of varying magnitude and frequency of landforming mass movements. Selby argued that in environments where denudation is dominated by mass movements, rapid changes in the state of landforms will be interspersed with periods of form adjustment. During these periods the landscape will contain erosional forms and deposits which are the result of the previous event. Thus the landscape bears evidence of past events (e.g. Figure 7.10). In

Figure 7.10: Hillslope form dominated by a past, high magnitude mass movement

some environments the landforming events will bring about major changes (high magnitude events) so that the historical aspect of landscape is

Form and Process

large. Elsewhere changes will be small (low magnitude events) with a correspondingly lower emphasis on history in the appearance of the landscape. Selby contended that the frequency of events may be independent of their magnitude so that in some environments large events can occur frequently and be the dominant landforming agencies, but elsewhere events of the same magnitude may be less frequent and less important as landforming agents (Figure 7.11). Hillslope forms differ in response to these differences in magnitude and frequency.

Figure 7.11: Hypothetical scheme for magnitude-frequency relations of landform-changing processes in different environments (*After Selby, 1974a, Engng. Geol.,9. Reproduced by permission*)

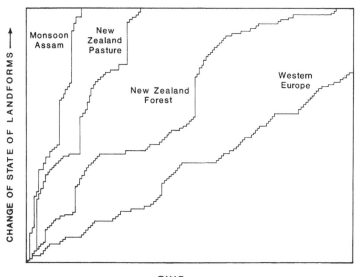

For wash processes some studies of magnitude and frequency have sought to establish which size of event causes most denudation (e.g. Pearce, 1976a), whereas others have given more attention to the qualitative differences between events of different magnitudes. Mills (1981), in a study in southwest Virginia, has argued that runoff from rainstorms with frequencies of 10^{-1} to 10^{-2} years is capable of incising gullies only where the underlying shale is not protected by a coarse

134

bouldery armour. Since this armour covers existing gully floors, rainstorms of this frequency lead to incision at the sides, and hence migration, of gullies. Rarer events (frequency of about 10^{-3} years) can generate sufficient runoff to entrain the bouldery deposits which are shifted into the gully floors. Thus concentrated erosion, as a result of gully migration, causes areal hillslope denudation as was proposed earlier by Bryan (1940) in his concept of *gully gravure*. The effects of runoff events of different magnitudes and frequencies have been incorporated into a simulation of hillslope development by Band (1985).

The second major difficulty in applying the concept of magnitude and frequency to hillslope processes lies in the implicit assumption by Wolman and Miller (1960) of stationary probabilities for events of given magnitude. It is assumed, for example, that the probability of a stream discharge of a given magnitude is constant over time. Such an assumption cannot be made for hillslope processes, particularly mass movements. Once a mass movement has occurred at a site, time may be required before conditions (e.g. accumulation of sufficient material) allow another such event (Innes, 1985). Furthermore, land-forming events at particular sites may well modify those sites to reduce the probability of further similar events at that site. Gretener (1967) argued that as denudation proceeds the probability of a major landslide will decrease and that landslides of great magnitude are restricted to young mountain chains. This argument lies behind Selby's (1974a) claim for different frequencies for events of given magnitude in different environments (Figure 7.11). Although the argument of the preparedness of the site may not apply to denudation by wash processes for which the frequency of events is determined by meteor-ological conditions, long-term landform change may nonetheless affect the landforming significance of particular rainstorm events. In so far as the concept of geomorphic thresholds (Schumm, 1973) allows separation of the probability of an event from its landforming effect, objections based upon preparedness of the site may be overruled. Nevertheless, progressive changes to hillslopes as a result of particular events do undermine the uniformitarian principle of Wolman and Miller's concept of magnitude and frequency when applied to hillslopes.

THE ROLE OF TIME: 8
EVOLUTION

For more than a century, indeed for as long as the study of hillslopes has been an identifiable branch of geomorphology, the way in which hillslope form changes through time has been a topic of investigation. Approaches to the issue have varied. An attempt to rationalise these approaches based upon the input of theory, quantification and empirical data is made in Figure 8.1. Studies of hillslope evolution that

Figure 8.1: Classification of studies of hillslope evolution

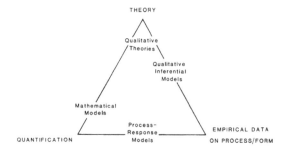

are described as pure theory are characterised by qualitative argument unsupported by any empirical data. These studies tend to be broad in concept and seek to provide a general model for hillslope evolution. Emphasis on empirical data often leads to studies that seek to explain hillslope evolution in a particular setting. Generalisations from such limited studies may or may not be attempted. An input of quantification leads to deductive mathematical models and process-response models. In the following sections several approaches to the study of hillslope evolution are examined. The inputs of theory, empirical data and quantification to each of these approaches can be seen from the positioning of the section headings in Figure 8.1.

QUALITATIVE THEORIES

Included within this group are some of the most widely known studies of hillslope evolution. The style of these studies is exemplified by the work of Davis (1899) who proposed an evolutionary model for valley sides within his wider concept of the geographical cycle. Davis's argument may be summarised as follows (Figure 8.2):

Figure 8.2: Davis's scheme for hillslope evolution

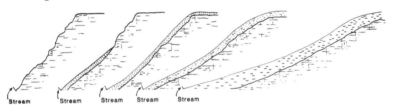

Bedrock surface

Coarse-grained waste sheet

Fine-grained waste sheet

(i) Valley sides are created by downcutting streams. Initially, the rocks through which the streams have cut are exposed at the surface of these valley sides and the detail of their morphology (cliffs and ledges) reflects the character of the subjacent rock.

(ii) Gradually, and working from the base upward, the valley sides are covered by a layer of weathered material which smooths out the morphological irregularities. This layer Davis termed the *graded waste sheet*. When the waste cover first develops it is coarse in texture, of moderate thickness and occupies steep gradients.

(iii) As the geographical cycle proceeds so the waste cover acquires a progressively finer texture and becomes deeper. At the same time

138

hillslope gradients become less, as the ever finer waste material can be transported on ever gentler gradients.

Although it is almost certain that Davis developed his evolutionary model for hillslopes based upon his own observations, he made no appeal to empirical data to support the model he proposed. Within the model only his assertion that the graded waste sheet increases in depth and fineness of texture as gradients decline is amenable to a form of empirical testing. Such tests were, however, neither carried out nor proposed. Davis's consideration of the processes leading to reduction in hillslope gradient through time was so vague that no empirical data collection could be envisaged. As for the most important element of the model, namely the change in hillslope form through time, clearly direct evidence is unavailable. Davis made no reference to the use of observed spatial differences in hillslope form as analogues for changes through time. His proposal for the manner of hillslope evolution rests entirely upon unsupported reasoning. It is a measure of the intuitive attractiveness of Davis's reasoning that his model obtained widespread acceptance.

The models of hillslope evolution proposed by Penck (1924) and King (1953) have commonly been identified as rivals to that of Davis (e.g. Young, 1972, p.25). This status is probably due to the fact that all three claim worldwide applicability but show an immediately obvious difference in the manner of hillslope evolution. Whereas Davis claimed that hillslopes evolve by a gradual reduction in their gradient, Penck proposed evolution through replacement from below of steeper gradients by gentler ones and King argued for parallel retreat of hillslopes. Penck's and King's models differ from Davis's in that Penck's is more quantitatively argued and King's makes greater appeal to empirical data for support.

Penck's model may be regarded as lying between the wholly qualitative argument of Davis and the strictly quantitative analysis of mathematical models (see below). Penck reasoned as follows:

(i) The effect of weathering on an exposed rock surface or regolith layer is to produce an ever finer-textured surface layer - a process termed *reduction*.

(ii) A surface layer has the property of

139

mobility. Mobility determines the gradient from which the layer can be removed by denudation.

(iii) Mobility is directly related to texture so that finer-textured surface layers are mobile on gentler gradients.

Penck applied this reasoning to an initial straight hillslope bounded at the top by a horizontal plateau and at the base by a stream which removes all material supplied to it but does not actively erode. The following evolutionary sequence emerges.

(i) The straight hillslope is uniformly exposed and subject to a uniform rate of weathering.

(ii) After time t, the surface layer attains sufficient mobility to be removed by denudation. However, a small fragment at the base of the hillslope is supported from beneath and is not removed.

(iii) If processes (i) and (ii) are repeated over several intervals of time t_1 to t_n an ever longer section develops at the base of the hillslope but inclined at a lower gradient. (Figure 8.3A).

(iv) The surface layer of this lower element continues to weather until it eventually acquires sufficient mobility to be removed at that gradient. The argument about the lowest fragment being supported from below applies here also, so that a further element inclined at a yet lower gradient develops (Figure 8.3B).

(v) If the time intervals t are made very small and the surface layers removed by denudation are made very thin, each hillslope element becomes very small, the difference in gradient between adjacent elements is also small and a smoothly concave profile results.

Penck's argument thus shows that hillslopes become flatter from the base upwards and that retreat of any part of a hillslope results in its lowest portion having its existing gradient replaced by a gentler one. In developing his model Penck defined a set of process rules which he then applied to an initial hillslope form. In this sense his approach is identical to that of mathematical and process-response modelling discussed below. The difference is that his argument is almost wholly qualitative. Penck made no attempt to found his process rules in empirical data on processes. Insofar as processes do not operate as he envisaged, (in particular,

weathering does not automatically lead an increase
in mobility) Penck's model is unrealistic.

Figure 8.3: Penck's scheme for hillslope evolution
  A.Parallel retreat of an initial hillslope and
     development of a lower-angle footslope
  B.Simultaneous parallel retreat of the initial
     hillslope and the lower-angle footslope
(*After Penck, 1924. Reproduced from the English
translation by H. Czech & K.C. Boswell, 1953,*
Morphological Analysis of Landforms *by permission of
Macmillan, publishers*)

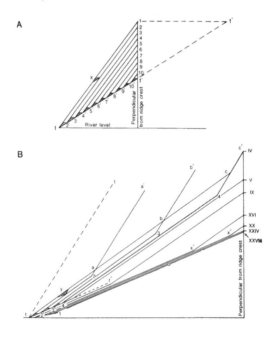

King's rejection of the Davisian model was
founded upon his claim that empirical evidence
contradicts Davis's conclusions. He cited plateau
landscapes of inland Africa of great antiquity but
with thin soil covers, retreat of the Drakensberg
scarp without any apparent reduction in gradient,
and the fact that large and small residual hills
have similar maximum gradients which he took to
imply that once a stable gradient is achieved this
gradient is maintained as residual hills are
further reduced in size. King's own model for

141

hillslope evolution is based upon an earlier model developed by Wood (1942) which, in turn, can be seen to incorporate significant elements from the model for evolution of desert hillslopes proposed by Lawson (1915).

Lawson considered the evolution of an upfaulted block mountain under a rainless climate. He argued that provided the initial gradient of the mountainside exceeds the angle of repose of detritus, the products of mechanical weathering will accumulate in a wedge of talus. As this talus layer grows the detritus from an ever-shortening mountain face is spread over the surface of an ever-lengthening talus slope. Consequently a greater amount of retreat of the mountain is required for a constant increase in the height of the talus wedge. Ultimately the whole mountainside will be covered by a rectilinear talus wedge overlying a convex bedrock surface (Figure 8.4). In a rainless climate no processes would transport away the talus material.

Figure 8.4: Lawson's scheme for hillslope evolution in a rainless climate. The mountain face OR is reduced to OPR_1, the lower curved portion of which OP would be buried in talus OSP. The combined rock and talus slope SPR_1 is the unchanging profile under the assumed conditions (*After Lawson, 1915*, Bull. Dept. Geol. Univ. California,9. *Reproduced by permission of University of California Press*)

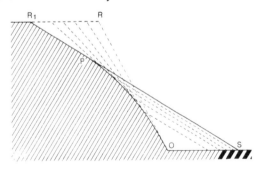

Wood extended this idea to consider the effects of transportational processes acting on the talus slope, which he termed the *constant slope*, as well as weathering of the initial bedrock surface, which he termed the *free face*. Wood argued that transportation of finer material from the constant slope would lead to a size sorting; coarser

142

material at the top, finer material at the bottom. This leads to a concavity at the base of the constant slope, termed the *waning slope*. In a climate with rain, processes other than weathering affect the free face leading to a rounding of its junction with the plateau surface above. Wood termed this convexity atop the free face the *waxing slope*. He proposed an evolution of hillslopes based upon the appearance and subsequent eradication of these four hillslope elements (Figure 8.5).

Figure 8.5: Wood's slope cycle. A. Free slope only, cut in flat recently upraised land surface. B and C. The constant slope forms. D and E. Waning slope develops. F. Waxing slope forms, waning slope rises up sides of constant slope, alluvial filling represented by dots. G. Constant slope has been consumed, alluvial fill deepens, slopes gradually flatten and approach a peneplane (*After Wood, 1942*, Proc. Geol. Ass.,53. *Reproduced by permission*)

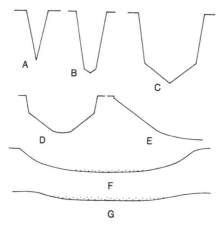

King elaborated upon the processes acting upon Wood's four hillslope elements and on the way in which these processes bring about retreat of the hillslope. The basis of King's model is that the free face retreats parallel to itself by weathering and rill erosion. As the free face retreats so do the other hillslope elements leaving an extending waning slope or *pediment* at the base. King claimed that water that flows in rills down the free face spreads out laterally on

143

the pediment and flows in a sheet of water, and that whereas flow in the rills is turbulent, sheetflow is laminar. In the absence of a free face, such as in areas of weak rocks or low relief, King claimed that hillslope evolution is extremely slow.

In contrast to those previously discussed, King's analysis of hillslope evolution makes reference to empirical measurements of hillslope form and refers in much greater detail to hillslope processes. However, it cannot claim to be soundly based in empirical data. In particular, no evidence beyond the author's own assertions is presented concerning hillslope processes. For example the claim that water flow changes from being turbulent to being laminar as it descends the hillslope is not verified.

Qualitative theories of hillslope evolution, such as those presented in this section, were proposed over a period of about fifty years ending soon after the middle of the present century. There can be little doubt that the growing body of empirical data on both hillslope form and processes will inhibit the emergence of further models of this type, based as they are on no more than assertion and reasoning.

QUALITATIVE INFERENTIAL MODELS

If it is possible to identify successors to the models discussed in the previous section, qualitative inferential models are those successors. The emphasis on reasoning remains but assertion is replaced by empirical data. An important difference between this group and the models of the previous section is that they are more limited in scope. They seek to explain hillslope evolution within a particular locality or under a closely defined set of conditions. All models employ measurements over a spatial set of hillslopes as a surrogate for measurements in a temporal sequence.

Savigear (1952) identified a section of marine cliff that had been progressively cut off from marine attack by an eastward growing sandspit. He was thus able to claim that the east-west set of hillslope profiles that he measured along this cliffline represented a temporal sequence (Figure 8.6). Savigear claimed that so long as the base of the cliff is subject

144

to unimpeded basal removal, hillslopes are modified by parallel retreat and the hillslope exhibits rectilinear segments that intersect at angular discontinuities. However, where there is impeded basal removal hillslopes develop smoother convex and concave forms and are modified by gradient decline. Although Savigear recognised that his conclusions applied only to the cliffline that he had measured, he suggested that conditions of impeded or unimpeded basal removal may be critical in determining whether hillslopes evolve by decline or parallel retreat.

Figure 8.6: Set of profiles measured along an abandoned cliffline in S. Wales. Profiles are arranged in order (A to N) of increasing time since they were subject to marine attack at the base (*After Savigear, 1952*, Inst, Br. Geogr. Trans.,18. *Reproduced by permission*)

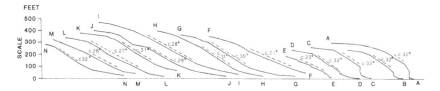

Other, later studies that have sought to identify spatial sets of hillslope profiles that may be interpreted as temporal sequences include those of Carter and Chorley (1961), Brunsden and Kesel (1973) and Dunkerley (1980). Carter and Chorley argued that, in a stream network undergoing headward extension, valley-side age and stream order are positively correlated. They examined the relations between stream order and mean and maximum valley-side gradient and found an increase in maximum gradient (up to order 4) and in mean gradient (up to order 3). With further increase in stream order valley-side gradient decreases. Carter and Chorley's study illustrates the complexity of factors that determine the manner of hillslope evolution. For their study area, it appears that hillslopes become steeper during the period that they are subject to basal stream attack and are increasing in total height. Once basal stream attack and downcutting ceases, hillslopes begin to decline in gradient.

Both Brunsden and Kesel and Dunkerley raised the question of absolute dating when using spatial

145

sets of profiles to infer temporal sequences. As Brunsden and Kesel pointed out, without the ability to date individual profiles or groups of profiles, there is the possibility that the observed variability in the spatial set represents no more than random variation of a time-independent form rather than an evolutionary sequence. Both studies present data in which closer dating of measured profiles can be achieved. In Dunkerley's study hillslope profiles were measured on four Pleistocene marine terraces in Papua New Guinea, ranging in age from 125,000 years to 277,000 years. The results show an increase with age in hillslope gradient but a coincident decrease in profile curvature. However, the significance of these findings is open to question. Profiles measured on younger terraces are not only gentler and more rectilinear but longer and have less height. To what extent the former two properties are controlled by the latter two rather than difference in age *per se* cannot be determined.

In other inferential studies, the relationship of hillslope profiles to geological structure (commonly a horizontal or near-horizontal caprock) has been used to infer a temporal sequence of hillslope profile form. The early study by Fair (1947) placed considerable emphasis on the role a dolerite caprock in controlling parallel retreat of the hillslope (Figure 8.7). Later studies

Figure 8.7: Evolution of a hillslope profile during and after the removal of a dolerite caprock. (*After Fair, 1947,* Trans. Geol. Soc. S. Africa.,50. *Reproduced by permission*)

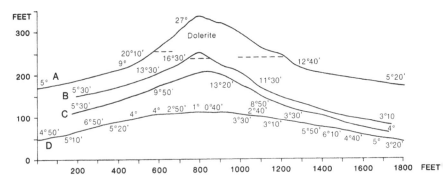

146

(e.g. Ollier and Tuddenham, 1962; Mills, 1978; Pain, 1986) have reached similar conclusions. Mills focused on the evolution of hillslopes once the caprock had disappeared. He demonstrated a linear relationship between maximum hillslope gradient and the vertical distance between the hilltop and the base of the former caprock (Figure 8.8). He attributed this relationship to the progressive fining of the surficial debris on the hillslopes as the hilltop is lowered. Mills speculated on the cause-effect relationship between fining of the surficial debris and reduction in sediment transport rate on the lower gradients.

Figure 8.8: Hillslope gradient plotted as a function of the vertical distance between the hill top and the base of the former caprock. Solid circles represent hills capped by Pennington Formation, and regression line is based on these points. Open squares represent steepest gradients measured on hills with intact caprocks. (*After Mills, 1978*, J. Tennessee Acad. Sci.,53. *Reproduced by permission of the author*)

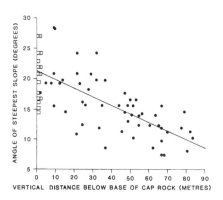

Mills's study is characteristic of the greater attention given to processes in more recent inferential models of hillslope evolution. It is no longer sufficient to infer a plausible evolutionary sequence of observed hillslope forms; the processes acting on the hillslopes need to be plausible also. Pain argued that in his study area the sandstone caprock induces landsliding in the underlying shale, thereby progressively undermining the caprock. The shale continues to

147

fail by mass movement until it approaches $10°-12°$, which gradient Dunkerley (1976) had previously shown to be the angle of long-term stability predicted from analysis of the residual shear strength of the shale regolith.

A difficulty of including observations of processes when inferring hillslope evolution lies in the uncertainty that present-day processes have been responsible for fashioning the observed hillslopes. In his study of hillslope evolution in part of Antarctica, Selby (1971, 1974b) was able to point to evidence indicating more than three million years of near uniform climatic conditions to justify his claim that the hillslopes of his study area had evolved under the dominant influence of the salt weathering that he observed. Smith (1981) was unable to present such convincing evidence for the processes fashioning the hillslopes he studied in Nigeria. He recognised that his proposal for the manner of hillslope evolution could be no more than a preliminary model and that further progress in understanding hillslope evolution would require detailed and prolonged measurements of hillslope processes. In coming to this conclusion, Smith pinpointed the weakness of the inferential approach to modelling hillslope evolution. Without a substantially greater input of data on processes, which in effect means replacing qualitative observations with measurements, and a certainty that the present-day processes are responsible for fashioning the hillslopes (see chapter 9), there is little more to be achieved by this approach to understanding hillslope evolution.

MATHEMATICAL MODELS

This distinctive approach to the question of hillslope evolution may be dated back to the work of Fisher (1866) who proposed an evolutionary path for a chalk cliff based upon deductive argument (Figure 8.9). Fisher's argument is essentially the same as that presented by Lawson almost half a century later. The important difference is its more mathematical approach which may be presented as follows. A rectilinear cliff of height h and gradient β is bounded at its upper and lower ends by horizontal surfaces. The cliff is uniformly exposed to weathering so that in unit time a layer of rock of uniform thickness falls away from it. This rock debris accumulates at the base as scree, with a rectilinear surface gradient α. Once buried by scree the former cliff is protected from further weathering. Just as Lawson (1915) later argued, rockfall from an ever shorter cliff is spread over an ever longer scree surface so that when, in time, the entire cliff is replaced by a rectilinear scree slope of inclination α it is underlain by a parabolic bedrock surface. Knowing h and α, and assuming $\beta = 90°$, the equation of the bedrock surface is given as

$$y^2 = 2h x \tan \alpha$$

Fisher's simple model was extended by Lehmann (1933) who considered the effects of varying β and then examined the ratio of the volume of removed rock to the volume of scree expressed as

$$\frac{rock\ volume}{scree\ volume} = \frac{1-c}{1}$$

where c is a constant. Fisher had assumed the two volumes to be equal (c=0) but, because of the voids in loose scree, this can be the case only if some of the material falling from the cliff face is removed and does not accumulate as scree. Fisher's parabolic shape for the buried bedrock surface is, in fact, a special case from a family of logarithmic curves, the individual members of which are determined by c.

Figure 8.9: Fisher's model of hillslope evolution. (*After Fisher, 1866*, Geol. Mag.,3)

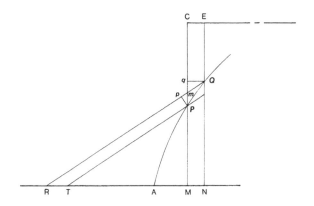

An important end-member of the family, given by c = - , was considered by Bakker and Le Heux (1952). In this case the bedrock surface is rectilinear and has gradient α. This condition is equated with the Richter slope named after the geomorphologist who identified bare rock surfaces in the Alps inclined at the gradients of scree slopes (Richter, 1901).

Other variants on Fisher's model have been presented by Bakker and Le Heux (1946, 1947, 1950), Van Dijk and Le Heux (1952) and Looman (1956). Among other things, these variants consider the effects on the shape of the buried rock surface of non-uniform rates of weathering of, and rockfall from, the cliff face. In particular, Bakker and Le Heux (1947) argued that the rate of weathering on the cliff face would increase with distance from the base of the cliff. Consequently, instead of a slice of uniform thickness being removed from the cliff in unit time (parallel rectilinear recession), they proposed that a wedge increasing in thickness from the base upward would be removed. They termed this mode of cliff retreat *central rectilinear recession*. A feature of the developments from Fisher's initial model was the presentation of the models in the more formal notation of differential equations giving the form of the bedrock surface as

$$\frac{dy}{dx} = \frac{h-cy}{h.\cot\beta -(\cot\alpha -c.\cot\alpha -\cot\beta)y}$$

The salient weaknesses of Fisher's model and its derivatives are threefold. First, it is assumed that weathering takes place only on the exposed cliff face. Processes are assumed to affect neither the buried rock surface nor the scree surface. Second, the assumption about the operation of processes is entirely without empirical foundation. Although Fisher's model was inspired by observation of a real field situation, it is developed from his interpretation of that situation. Other interpretations are no more and no less valid. Thus there is no basis for preferring Bakker and Le Heux's notion of central rectilinear recession over Fisher's of parallel rectilinear recession. Third, with the exception of the case where $c = -\infty$, the model is concerned with the shape of a bedrock core beneath a wedge of debris. In all cases the ground surface is assumed to have a rectilinear form inclined at the angle of stability of the debris. Recognition of these weaknesses has led to developments in this type of model of hillslope evolution.

Scheidegger (1961b) has considered the manner of hillslope evolution where $c = -\infty$ under a variety of assumptions about the relationship of the rate of denudation to the shape of the hillslope. Under the first group of assumptions, termed *linear theory*, the rate of change of elevation of a point on the hillslope is a constant

$$\frac{\delta y}{\delta t} = -\text{const}\Phi$$

In this group Scheidegger considered three cases (Figure 8.10), as follows

case 1: $\Phi = 1$
case 2: $\Phi = y$
case 3: $\Phi = \delta y/\delta x$

Figure 8.10: Linear theory of hillslope evolution (*After Scheidegger, 1961a*, Geol. Soc. Am. Bull.,72, 40-1. *Reproduced by permission*)

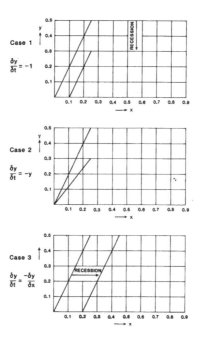

151

Scheidegger argued that linear theory suffered from the serious oversimplification of assuming that it is vertical lowering that is related to the constant Φ. In reality weathering is normal to the ground surface so that it is more realistic to assume that vertical lowering can be represented by the equation

$$\frac{\delta y}{\delta t} = -\sqrt{1 + (\frac{\delta y}{\delta x})^2}\ \Phi \qquad (8.1)$$

Because this differential equation is non-linear, he termed the model based on this assumption *non-linear theory*. Unlike the equations of the linear theory, easy solutions of equation 8.1 do not exist but approximate solutions can be obtained using difference equations. For the same three cases as considered in the first group, solutions to the difference equations lead to hillslope evolution as shown in Figure 8.11. Scheidegger claimed, on the basis of closer agreement with hillslopes in nature, that case 3 is the most reasonable one and elaborated on this model to consider the effects of basal undercutting and variations in rock resistance (Scheidegger 1960, 1964).

Ahnert (1964) argued that it is neither realistic to assume that the debris produced by weathering is instantaneously removed nor to expect that its accumulation prevents further weathering of the rock beneath it. On the contrary, field experience indicates that most hillslopes bear waste covers yet bedrock weathering continues beneath them. Ahnert developed models of hillslope evolution based on the postulates that the rate of weathering is a function of the thickness of the overlying waste mantle and that the waste mantle moves slowly downslope at a velocity proportional to the sine of hillslope gradient. These models, representative of those being developed by several workers at that time, extend Fisher's basic idea in three fundamental ways. First, whereas Fisher was concerned only with weathering of cliff faces, these models consider weathering of rock inclined at any angle. Indeed, some studies considered only rock surfaces inclined below the angle of repose of weathered debris. Second, Fisher assumed weathering debris to fall off cliffs instantaneously. His model takes no account of rock surfaces inclined below the angle of repose of debris but Lawson (1915) assumed that it would

Figure 8.11: Non-linear theory of hillslope evolution (*After Scheidegger, 1961a*, Geol. Soc. Am. Bull.,72, 43-5. *Reproduced by permission*)

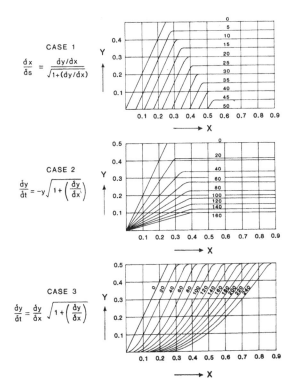

remain in place to form a protective layer preventing further weathering of the rock. These models assume weathered material will be removed from gentler hillslopes, albeit slowly. Third, Fisher gave no further consideration to the accumulating debris. In these later models accumulating debris is assumed to be as subject to the forces causing downslope movement as freshly weathered debris.

A number of studies illustrate this type of mathematical model of hillslope evolution. Culling (1963) presented a diffusivity-type equation,

$$\frac{\delta y}{\delta t} = K \frac{\delta^2 y}{\delta x^2}, \qquad (8.2)$$

153

Evolution

where K is the coefficient of diffusion. This model is applicable to evolution of hillslopes that satisfy two conditions: (1) that the rate of removal of weathered debris is proportional to hillslope gradient, and (2) that there is always weathered material available for transport, i.e. the hillslope is transport-limited. Young (1963) presented a series of models based upon a variety of assumptions about the rate of removal of debris, the relationship of weathering to debris cover thickness, and the conditions at the base of the hillslope. Souchez (1963, 1966) developed a model for evolution of a hillslope under the influence of transport-limited mass movements based on the relationship of velocity of downslope movement v to tangential shear stress τ, critical shear stress τ_c and viscosity η expressed as:

$$\frac{-dv}{dy} = \frac{\tau - \tau_c}{\eta}$$

Many of the ideas included in the work of Scheidegger and Culling are incorporated in the studies of hillslope evolution presented by Hirano (1966, 1968, 1972b, 1975, 1976) who combined cases 2 and 3 of Scheidegger's linear theory with Culling's diffusion model to yield

$$\frac{\delta y}{\delta t} = a \frac{\delta^2 y}{\delta x^2} - b \frac{\delta y}{\delta x} - cu$$

in which the constants a, b and c are termed the denudational coefficient, the recessional coefficient and the subduing coefficient, respectively. In a series of analyses Hirano examined the consequences for hillslope evolution of a variety of values of these three coefficients as well as considering the effects on hillslope evolution of changes in hillslope basal condition.

Although other studies could be discussed (e.g. Luke 1972, 1976; Carter and Nobes, 1980) and although each has its own character, all models discussed in this section have a common weakness. The equations which determine the manner of transformation of some initial hillslope form at best are weakly founded upon intuitive generalisations about the operation of hillslope processes, and at worst are wholly arbitrary. Firmer grounding of mathematical modelling of hillslope evolution in empirical data on the operation of hillslope processes characterises process-response models of hillslope evolution.

154

PROCESS-RESPONSE MODELS

The distinction between the models described here as process-response models and those of the previous section is not clearcut. Hence, Young (1972, pp.109-16) described as process-response models many discussed in the previous section. In the widest sense any model that applies an operator (usually expressed as a mathematical function of x and y) to an initial form (usually defined by its coordinates in x and y) is a process-response model. However, in the present survey the classification is restricted to those models in which the operator is clearly based upon empirically derived, quantitative data on hillslope processes, and hence is a real entity (Whitten, 1964).

The majority of process-response models employ simulation techniques to identify hillslope evolution. A smaller number have adopted analytical methods. Kirkby (1971) considered the case of hillslopes evolving under transport-limited processes where the capacity of the processes could be expressed as:

$$C = f(x) \cdot \left(-\frac{\delta y}{\delta x}\right)^n, \tag{8.3}$$

where n is a constant. Assuming that contours are straight so that there is no transport of sediment into or out of individual profile lines, changes to a hillslope profile through time t can be expressed by a continuity equation of the form:

$$\frac{\delta C}{\delta x} = -\frac{\delta y}{\delta t}, \tag{8.4}$$

For n=1 and assuming that the base of the hillslope remains fixed through time ($y_1(t) = 0$), there exists a solution to the continuity equation of the form:

$$y = U(x) + V(x) \cdot T(t),$$

where U,V are functions of x alone, and T is a decreasing function of t alone. This solution describes the form, the *characteristic form*, to which the hillslope profile tends as time passes, irrespective of the initial form of the hillslope profile. For values of $n \neq 1$ it is more difficult to obtain an exact solution to equation 8.4 (Trofimov and Moskovkin, 1983) but Kirkby assumed that similar solutions can be found for other values of n to yield a family of characteristic

155

Evolution

forms. Figure 8.12 shows characteristic forms with values for n of 1 and 2 and, assuming f(x) to have the form x^m, with values of m lying between 0 and 2.5. Herein lies the weakness of the analytical method. In order to obtain solutions to equations it is frequently necessary to make

Figure 8.12: Characteristic hillslope forms for a range of processes, the transporting capacities of which can be represented by equations of the form of equation 8.3 and where f(x) has the form x^m. (*After Kirkby, 1971*, Inst. Br. Geogr. Special Pubn.,3. *Reproduced by permission*)

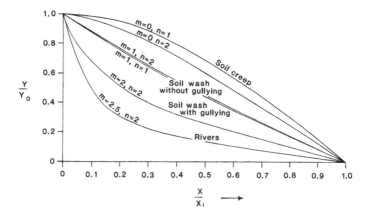

assumptions, simplifications or to seek approximations. As understanding of the operation of hillslope processes becomes more detailed, the expression of these processes in equations of the form of 8.3 will become more difficult and the likelihood of being able to solve these equations analytically will become proportionately less. Furthermore, as Kirkby (1976b) has pointed out, such process-response models are both deterministic and continuous. They predict a single hillslope form resulting from the operation of processes and they can model only processes that operate continuously. In reality, all hillslope processes are discontinuous. Even a process such as soil creep is made up of a very large number of individual heave movements, the magnitude of each of which depends, among other things, on the sequence of meteorological events. The same process may operate on two hillslopes but the effects may differ simply in response to local

156

variations in weather conditions. To be more successful process-response models need to take account of the discontinuous nature of hillslope processes and to predict not only the hillslope form that results from the operation of these processes but the likely variability that will be exhibited by that form.

Numerical simulations of hillslope evolution employ the same basic continuity equation but, rather than seeking analytical solutions to the equation, they monitor the changes at specific points on a hillslope through time, as follows. An initial form is defined by the coordinates of the specified points. An equation of the form of 8.3 defines the rate of sediment transport at each point. The quantity of sediment transported past each point in unit time is calculated and, using a form of the continuity equation, the changes to the elevations of these points are determined. These new coordinates then form the input hillslope for the next iteration. The procedure is repeated for as many iterations as desired. A number of studies have utilised this approach to model evolution of hillslope profiles (e.g., Ahnert, 1973; Parsons, 1976b; Kirkby, 1976b, 1984, 1985; Trofimov and Moskovkin, 1984; Band, 1985). If equation 8.3 is used to define sediment transport the simulation will yield the same result as an analytical solution. The value in simulation lies in its greater ability to cope with more complex processes equations. Band incorporated into his version of the continuity equation a parameter that allowed modelling of fluctuations in precipitation and runoff conditions to permit analysis of the contributions to total sediment transport and hillslope development made by various magnitudes of precipitation and runoff intensity. He then converted the continuity equation to a finite difference form and obtained finite difference equations for the set of points defining the hillslope profile.

A significant feature of the models of hillslope evolution discussed so far is that they have all been concerned, explicitly or implicitly, only with the evolution of profiles on plan planar hillslopes. Clearly this is a major shortcoming. Many hillslopes are curved in plan and during their evolution many profiles can be expected to gain or lose sediment as a result of plan concavity or convexity along the profile line.

Consideration of the problem of plan curvature was given by Kirkby (1971) in his analytical solutions to continuity equations, but it has been only in simulation studies that the question of evolution of hillslopes, rather than hillslope profiles, has been addressed. Ahnert (1976b) presented a model to allow simulation of weathering, surface wash, viscous flow, plastic flow and debris slides on an initial surface of any configuration and with basal conditions that could be varied during the simulation. Armstrong (1976) modelled the effects of weathering, soil creep and wash processes on an initial surface incised by streams (Figure 8.13).

Figure 8.13: Three-dimensional simulation of hillslope evolution. (*After Armstrong, 1976*, Zeit. für Geomorph. Supplementband,25. *Reproduced by permission*)

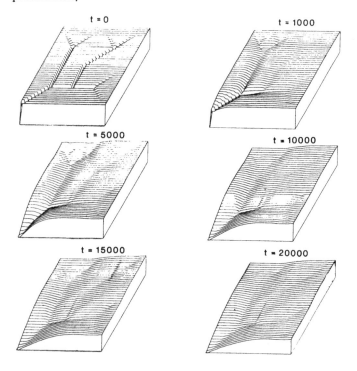

In a sense, extending hillslope process-response models into three dimensions has necessitated a step backwards towards mathematical models. These three-dimensional models demand data on the process of lateral migration of stream channels which is not available (Smith and Bretherton, 1972). Armstrong (1976) recognised this shortcoming and simply assumed that the stream channels remained fixed throughout simulation. Kirkby (1986) avoided the problem by defining the initial topography as a pseudo-fractal surface and allowing stream channels to form and be destroyed during simulation in response to the operation of hillslope processes. A further weakness of simulation models to date has been the limited number of data points on which they operate. Ahnert used a square grid of 10 by 10 points, Armstrong one of 40 by 40 and Kirkby (1986) one of 64 by 64. Even in this last case a cell size of 10 m by 10 m was considered necessary to simulate over an adequate total area. Inasmuch as properties of processes were calculated for a cell as a whole, cell size determined, for example, the minimum width of overland flow that could be modelled. A value of 10 m is clearly unrealistic. On natural hillslopes local concentrations of flow in rills are much narrower than this so that the simulation fails to model adequately the transporting capacity of such flows.

There can be no doubt that three-dimensional simulation models have great potential for studying hillslope evolution. The most recent data on process rates and mechanisms can be readily incorporated into such models, they can be designed to include magnitude-frequency data (Band, 1985) and they can be made to yield probabilistic results. An important requirement, however, is that they simulate processes at a realistic scale. A cell size of no more than 1 m by 1 m is probably necessary.

EVALUATING MODELS OF HILLSLOPE EVOLUTION

Whichever type of model is used to examine hillslope evolution, the question of evaluating its performance arises. Most model-builders have been remarkably shy of this question. For even the most quantitative models, the evaluation has often consisted of no more than a degree of satisfaction that the predicted hillslopes bear qualitative resemblance to actual hillslope forms (e.g. Scheidegger, 1961a, p.143; Kirkby, 1971, p.25). Yet evaluation of evolutionary models is vital if the activity of modelling is to be more than an interesting exercise and is to have a real input to understanding the geomorphology of hillslopes.

Only in the case of rapidly evolving hillslopes is it likely that the actual evolution of hillslopes can be compared with that predicted by a model. Using early photography of the site, Band (1985) was able to identify initial shapes of hillslope profiles in an abandoned gold mine and to compare the changes that had occurred in a period of approximately 100 years with those that were predicted from his process-response models. Even in this case it is not the evolutionary sequence that was compared but simply start and end points. Band noted that it was difficult to make any evaluation of his model because the initial form was not known with sufficient accuracy. In any case, because coincidence of the predictions of the model with the real present-day form of the hillslopes was used to calibrate the model, there can be little certainty that the real hillslopes evolved in the same way as the model indicated. Hence, it would be unwise to use the model to predict future change.

In a few instances spatial sets of hillslope profiles have been taken to be equivalent to an evolutionary sequence and this has allowed closer comparison of predicted forms with actual ones. Hirano (1968), Parsons (1976b) and Kirkby (1984) have all used the set of profiles measured by Savigear (1952) against which to compare the results of modelling. In all cases, however, the comparison was qualitative. Nash (1980b) measured three sets of hillslope profiles on wave-cut bluffs along the eastern coast of Lake Michigan. These sets were related, in order of increasing age, to the shorelines of the present lake, the Nipissing Great Lakes (3,700 to 4,000 BP) and Glacial Lake Algonquin (c.10,500 BP). Nash demonstrated statistically significant differences in the morphology of the three groups of profiles which he attempted to explain as a function of the operation of evolutionary processes that could be modelled by equation 8.2. By using the known age differences between sets of profiles, Nash was able to derive a value for the constant K and to predict evolutionary differences in hillslope form very close to the differences observed among the three sets of profiles.

Both Caine (1969) and Statham (1976) compared

160

field measurements of talus slopes with the forms predicted from models. The two authors used different models, each appropriate to the situation from which his field measurements were obtained, and both report close agreement, again qualitatively assessed, between actual and predicted forms. Statham extended his comparison to include consideration of particle sorting of the talus slopes. Many models of evolution predict characteristics of hillslopes other than their forms and these characteristics are equally useful for testing models. Ahnert (1970a) compared waste cover thickness on hillslopes with that predicted from a mathematical model and obtained very close agreement (Figure 8.14).

Figure 8.14: Comparison of waste cover thickness on actual hillslopes with that predicted from a theoretical model (*After Ahnert, 1970a,* Zeit. für Geomorph. Supplementband,9. *Reproduced by permission)*

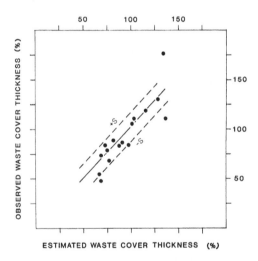

In a somewhat less well controlled experiment Pitty (1972) attempted to compare actual hillslope forms with those predicted by Davis's and Penck's evolutionary models as illustrated by Davis (1932, Figure 4, p.409). Pitty found closest agreement between actual hillslopes and those predicted by the two models in areas of homogeneous, non-cohesive rocks.

Models of hillslope evolution can be divided

161

into those that predict continuous evolutionary change in hillslope form (e.g. Davis, 1899) and those that predict an initial period of rapid morphological change followed by one of time-independent characteristic form (e.g. Kirkby, 1971). The latter type of model would lead to the expectation that the majority of naturally occurring hillslopes would correspond to characteristic forms, whereas the former predicts that hillslopes would exist at all stages in an evolutionary sequence.

More rigorous procedures for testing the predictions of models of hillslope evolution are necessary but it is not immediately apparent what they should be tested against. Moon (1977a) has commented that it would be unreasonable to compare simulated hillslopes with a single natural hillslope and that it is necessary to derive some representative against which to test model predictions. Moon presented a procedure for determining a representative hillslope profile from field measurements and used this procedure (Moon 1977b) to test the predictions from Ahnert's (1973) model. The difficulty of such comparisons lies in finding field situations that satisfy the conditions of the model. As simulation models become more flexible it may become possible to fit the conditions of the model to the field situation rather than vice versa. Three-dimensional simulations will allow measurements of sets of hillslope profiles for comparison with sets of field measurements and thus incorporate some of the natural variability in hillslope profile form.

Two problems bedevil all attempts to evaluate models of hillslope evolution. The more general of these is that of convergence and equifinality. Even if observed hillslope forms match those predicted by an evolutionary model there is no certainty that the forms were produced in the manner proposed under the model. Process-response models of the effects of both rainsplash and soil creep, for example, yield convex hillslope profiles. The existence of convex hillslope profiles is not a sufficient condition to argue that one or either of these processes has operated. This issue most severely affects process-response models, which attempt to relate hillslope forms to real processes, and all qualitative models that account for observed hillslope forms in terms of a particular sequence of events. The issue is least important for mathematical models which may predict hillslope forms that would result if processes operated in a manner described by particular equations. Inasmuch as these models need not specify the physical processes *per se* nor a particular sequence of events, they remain valid even if several processes can be accounted for by the same equations.

The second problem is that of inheritance. Even if it can be shown that particular processes are presently operating on hillslopes and process-response models of these processes yield hillslope forms like the observed ones, it cannot be assumed that these same processes have acted throughout the period of time over which the hillslope has reached its present form. This issue is discussed at greater length in the next chapter.

OBSERVATIONS OF REAL CHANGES THROUGH TIME

Very occasionally geomorphologists have the opportunity to study hillslope evolution not through models but by direct observation. Such opportunities arise in small-scale landforms and those composed of very easily eroded materials. Haigh (1979) monitored erosion on spoil mounds for a five-year period and was able to identify

Figure 8.15: Model for the evolution of abandoned spoil banks based upon observations of hillslopes of 60-years' and 30-years' age and measurements of hillslope processes over a 2-year period (*After Goodman & Haigh, 1981,* Physical Geography,2. *reproduced by permission*)

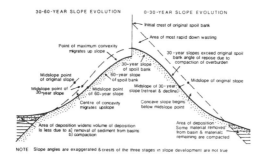

163

significant variations in sediment loss along profiles. In a similar study, Goodman and Haigh (1981) used morphological differences between spoil banks of different ages, together with measurements of erosion from them over a two-year period, to identify the manner of their evolution (Figure 8.15). The significance of such studies for evolution that takes place in a much longer time period is open to question but they are of interest in themselves and may have considerable value for predicting the short-term behaviour of many man-made hillslopes.

THE ROLE OF TIME: 9
INHERITANCE

How old are the hillslopes we see around us? This question is fundamental to all attempts to explain the diversity of hillslope form. Without a timescale for the evolution of hillslopes under the action of particular processes and an awareness of the length of time that hillslopes persist in the landscape after their formative processes are no longer active, it is impossible to establish relationships between hillslope form and the factors which control it. These issues and their consequences are examined in this chapter.

RATES OF HILLSLOPE EVOLUTION

Several studies have attempted to measure rates of denudation in such a way that they can be converted to rates of hillslope degradation. In reviewing these, Saunders and Young (1983) indicated that rates of 0.01 - 0.1 mm/yr and 0.1 - 1.0 mm/yr may be typical for normal and steep relief, respectively. Taking the median values from these two ranges (0.05 mm/yr and 0.5 mm/yr) and assuming an initial relief in both cases of 100 m (likely to be a significant underestimate in the case of steep relief) it would take 500,000 years and 50,000 years, respectively, for the initial relief to be reduced by half. Kirkby (1971, p.22) demonstrated that it is at this stage in land degradation that his process-response models yield characteristic forms. Thus, assuming the validity of Kirkby's models, it can be argued that for the majority of environments hillslope process-form relationships will take time of these orders to be established. Possibly only in badlands and anthropogenic landforms created of unconsolidated materials is it reasonable to expect that characteristic forms will be very much more rapidly established. In these environments rapid rates of degradation (upwards of 1 mm/yr) coupled with small initial relief (of the order of 10 m) may lead to characteristic forms in a few hundred to a few thousand years.

The implications of these rates of hillslope evolution are threefold. Firstly, it can be anticipated that relict hillslope forms will be common where substantial changes in hillslope processes have occurred during the last 10,000 to 100,000 years. Secondly, the morphological effects of intermittent, though currently active, processes will persist for substantial lengths of time. Thirdly, and less obviously, processes that operated in the past may act to constrain or modify those presently active.

RELICT HILLSLOPE FORMS

Büdel (1977) has claimed that all hillslopes in mid-latitudes having gradient less than 27° to 30° owe their morphology to Würm periglacial conditions and have survived almost unchanged since that time. Even if such an extreme position cannot be maintained, there is evidence that some hillslope forms must exist largely as relics from former climatic conditions. Examples of these features are the talus slopes which occur widely in upland Britain. For the most part, however, these accumulation forms are vegetated which suggests that rockfall is no longer a significant process (Figure 9.1). Two such talus slopes were investigated by Ballantyne and Eckford (1984). On the basis of short-term measurements of rockfall activity these authors inferred rates of talus accumulation of 0.004 mm/yr and 0.001 mm/yr which they concluded to be negligible in comparison with present rates of erosion of the talus slopes. In contrast, however, the estimated rates of rockfall at the time of the Loch Lomond Stadial are at least two orders of magnitude greater (Sutherland, Ballantyne and Walker, 1984). On this basis Ballantyne and Eckford concluded that such talus slopes formed in the interval between ice-sheet deglaciation and the end of the Loch Lomond Stadial c.10,000 B.P. The net erosion that has occurred during the succeeding period has been insufficient to eliminate depositional hillslope forms which, to a large extent, survive as relics from a long-past, process environment.

Figure 9.1: Relict talus slopes. Growth of vegetation cover indicates that rockfall is no longer very active

PERSISTENCE OF HILLSLOPE FORMS

Even where particular processes continue to be active, hillslopes commonly display a record of individual process events that may be dated back for several centuries. This is particularly the case for rapid mass movements in which large volumes of material are transported during short periods. Using lichenometry, Innes (1983) documented debris flows in Scotland, one of which he dated as having occurred in 1390. Hupp (1984) used dendrochronology to obtain dates of debris flows as much as 300 years old. In general, the larger the volume of material moved in a single event the longer its effects persist. A major event, such as the Blackhawk landslide in California, continues to have pronounced influence on hillslope form despite its antiquity.

Depositional features continue to exist on hillslopes until later or contemporaneous erosional processes are able to remove them. Major erosional events also create hillslope features that may persist as testament to their activity. In an examination of the effects of a large flood in the Cannon Hill valley, Exmoor, Anderson and Calver (1977) found that after little more than 20 years erosional features had suffered more degradational modification than had depositional ones, and that some mass movement scars were almost insignificant. There is a paucity of data concerning the persistence of erosional features on hillslopes. In addition, many of them have associated deposits so that it is difficult to tell how far a process event would be marked by the survival of erosion scars *per se* in the absence of depositional features.

LEGACIES FROM FORMER PROCESS ENVIRONMENTS

The magnitude of climatic changes over the last few million years when compared to estimates of the rates of hillslope evolution undermines attempts to identify simple relationships of hillslope form to process environments for all but the most rapidly evolving hillslopes. Furthermore, it implies that most hillslopes will owe part of their morphology to legacies from former climates. The criticisms made by Dunkerley (1978) of Toy's (1977) analysis of hillslope form and climate are, in part, concerned with this issue. Dunkerley pointed to the difficulty of establishing the climatic phase to which the evolution of Toy's sampled hillslopes belonged and commented that at some of the sites the discrepancy between the present and the Pleistocene climates is comparable to that currently observed along the length of Toy's traverses.

Legacies of former climatic conditions may be expressed not simply as survivals from these former conditions but as forms that would not have

168

existed but for a particular sequence of climatic change. Oberlander (1972) has documented the case of the boulder-strewn, granitic hillslopes of the Mojave Desert (Figure 9.2). He has argued that current weathering processes are not renewing the boulder mantles which are presently being disintegrated. Even though some of these hillslopes are in equilibrium with present-day processes (Parsons and Abrahams, 1987), their current morphology would neither have been brought about by processes active under the present climatic conditions nor by those of the late Tertiary, acting alone. The sequence of climatic changes is crucial to the existence of these hillslopes. Similarly, Goudie and Bull (1984) have accounted for characteristics of colluvial deposits on hillslopes in Swaziland by a process of corestone development within a deeply weathered layer and the subsequent stripping of the fines to leave the corestones at the surface.

Figure 9.2: Boulder-strewn hillslope in the Mojave Desert. The boulders are a legacy from a deep weathering mantle of a former climate. Current weathering processes are leading to the disintegration of the boulders but not renewal of the boulder mantle

Elsewhere, hillslopes may be complex assemblages of some elements that have survived from former climatic conditions amidst others that have developed under the current climate. Talus flat-irons (Figure 9.3) are commonly interpreted as such forms. Talus slopes develop under sub-humid climatic conditions but as these conditions are replaced by a more arid climate dissection and removal of the talus begins. Talus is stripped from the upper part of the hillslope first leaving the lower talus flat-iron as a remnant from the former climate. Such forms have been described from many environments including the southwestern United States (Koons, 1955), Cyprus (Everard, 1964) and southern Morocco (Schmidt, 1987).

Figure 9.3: Talus flat-irons. On the lower part of the hillslope the underlying rock is buried beneath a talus cover; on the upper part the former talus layer has been removed. In-facing scarps at the upslope end of the talus layer between the gullies that are presently dissecting it create the flat-iron shapes

THE AGE OF HILLSLOPES?

In seeking the answer to the question posed at the beginning of this chapter, we aim to determine the timing of the events that are responsible for the forms of hillslopes as they exist at the present time. If hillslopes have not been in existence for sufficient time for processes to have produced characteristic forms, then we need to look elsewhere to find the factors responsible for their present morphology. In some instances the initial configuration of hillslopes as a result of tectonic processes may contribute significantly to present-day morphology. Colman and Watson (1983) have remarked upon the different initial configurations that may be characteristic of different types of scarps. Similarly, rapid downcutting by streams in the past may be largely responsible for present-day valley-side form. Rates of cliff retreat in the eastern Grand Canyon as determined by packrat middens (Cole and Mayer, 1982) would indicate such a control.

There is no simple answer to the question of the age of hillslopes. In all environments it can be anticipated that close juxtaposition of hillslopes of very different ages may exist. The range of ages may be strongly influenced by the rates of denudation that apply to the particular environment and by the material of which the hillslopes are composed. However, these factors do no more than define the timescale appropriate to a particular hillslope setting. Theories of landscape evolution need to direct greater attention to the issue of the range in hillslope age than has been the case hitherto. Young (1983) has noted that "until now landforms of great antiquity have been regarded as puzzles in the matching of observation and theory" and has called for a different perspective to emphasise "the importance of factors glossed over in widely accepted evolutionary models".

THE INFLUENCE OF MAN

10

Man influences hillslope form in three principal ways. Firstly, and most obviously, he creates artificial hillslopes such as those of road cuttings and embankments and those in quarries. Secondly by effecting land-use change, man alters the balance of processes acting on natural hillslopes which leads to modifications in their form. Finally, in recent times man has sought to ameliorate some of the unsightly results of extractive industries by creating artificial landscapes.

ARTIFICIAL HILLSLOPES

Probably the most obvious characteristic of hillslopes created by man is their steepness. It applies equally to hillslopes created by cutting into natural landscapes (road cuttings, quarry faces and the like) as it does to ones created by accumulation of material (road embankments and spoil heaps). As Brawner (1966) pointed out, cost savings are possible if such hillslopes can be steepened. For quarry faces two considerations apply. The steeper the hillslope and the closer it is placed to the boundary of the land owned by the quarry, the greater the economic returns from the quarry. Brawner gave as an example the effect of increasing the gradient from 40° to 45° for a quarry face 2,000 feet long and 500 feet deep. This alteration in gradient changes the quantity of excavated material by 3,400,000 tons. Weighed against this benefit of steeper gradient is the increased chance of mass movements. A mass movement may endanger workers in the quarry, may damage machinery and is costly to clear away. In addition, there is the possibility that the mass movement may affect land adjacent to the quarry and owned by others. This is a particular problem for sand and gravel excavations which are often located close to population centres.

Mass movements are also a major hazard of the waste mounds associated with extractive industries. As Nir (1983) has pointed out, "the dumping will be carried out as cheaply as possible, the frictional forces of the usually loose material may be quite low, yielding a precarious equilibrium prone to disrupture by environmental factors". Again, the occurrence of such waste mounds close to centres of population signifies the need for safe design of these artificial hillslopes.

Safe design can be equated with stability. In many environments, long-term stability may be assessed from the form of natural hillslopes developed in the same material as the artificial ones. Elsewhere, suitable natural hillslopes with which to compare those created by man may not exist. Kirkland and Armstrong (1982) considered the case of road embankments on the Blackland Prairie, Texas. In this locality natural gradients are very gentle and give no guide to the limiting angle of stability. Kirkland and Armstrong pointed to slope failures on road embankments as evidence of excessive gradients and recommended that such embankments be constructed using residual strength values.

A further consideration in the design of artificial hillslopes is the length of time over which such a hillslope needs to be stable. Clearly, a quarry face needs to be stable for only a short period whereas longer-term stability is necessary for road cuttings and spoil banks. Abandoned quarry faces will evolve to attain gradients closer to those of long-term stability. Gagan and Gunn (1987) examined rockfall from limestone quarry faces near Buxton, Derbyshire. Their results showed that the most dramatic change in the form of quarry faces occurred in the 5 to 10 years post-abandonment. They also noted differing rates of morphological change in relation to methods of drilling and blasting employed during excavation. Uses that may subsequently be made of abandoned quarries and the timescales over which these new uses are implemented are significant for planning regulations in respect of such sites. At present there is probably inadequate data for firm guidelines to be produced. Further work on the rates of morphological change in artificial hillslopes, as well as studies of their sediment yield (e.g. Diseker and Sheridan, 1971), is required.

THE EFFECTS OF LAND-USE CHANGE

In broad terms, the effect of man-induced land-use change has been to increase erosion rates.

Figure 10.1: Changes in a drainage basin consequent upon land-use change. Erosion on the upper parts of hillslopes and aggradation of the lower parts causes upslope migration of the steeper part of the hillslope and an increase in maximum hillslope gradient (*After Strahler, 1956b*, in Man's Role in Changing the Face of the Earth, *W.L. Thomas ed. Copyright 1956 University of Chicago. Reproduced by permission of the University of Chicago Press*)

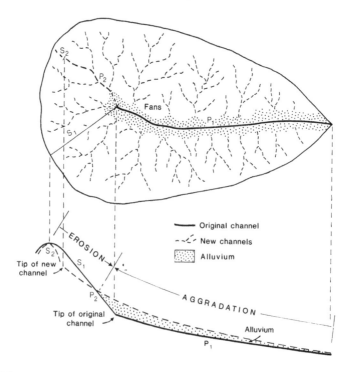

Strahler (1956b) has indicated the nature of the transformation of a drainage basin brought about by such an increase in erosion rates (Figure 10.1). Several studies have attempted to quantify the increase in erosion by overland flow associated with particular land-use changes. Meade (1982) estimated that erosion was increased by a factor of 10 consequent upon European settlement in North America. Richardson (1982) measured a more than 6-fold increase in soil erosion in Jamaica as a result of replacing natural rain forest with tropical pine forest. Robinson and Blyth (1982) estimated that in the short term sediment losses were increased more than fourfold as a result of drainage undertaken in connection with afforestation of a Pennine moorland. An extreme case of almost total destruction of vegetation cover near Sudbury, Ontario was investigated by Pearce (1976b). Changes in hydrologic regime were found to be associated with an increase in denudation rate of two orders of magnitude.

Increases in overland flow rates have two effects on hillslope form. In the short term, because streams are unable to evacuate their catchments of the increased supply of sediment, there is accumulation of sediment on the lower parts of valley sides accompanying the erosion from the upper parts. In consequence, hillslopes may become more sharply concave in profile. The second effect of increased overland flow is to enlarge the drainage network. In some instances this is expressed by numerous shallow rills (Figure 10.2) but, as Tacconi, Billi and Montani (1982) pointed out, rills are an embryonic form of extending drainage streams whose long-term equilibrium form is a network of smaller streams and shorter valley sides (Figure 10.3).

Overland flow is not the only process to be affected by land-use change. Several studies have reported increased frequency of mass movements consequent upon land-use change. Dating of debris-flow deposits in Scotland led Innes (1983) to the conclusion that land-use effects, particularly burning and overgrazing, have been responsible for the increased frequency of debris flows during the last 250 years. In more recent times, clearcut logging has been identified as a cause of increased landslide frequency (Ziemer, 1981; Rice and Pillsbury, 1982). The increased occurrence of landslide scars and deposits has an

Figure 10.2: Hillslope rills.   One   form   of enlargement of the drainage network

obvious effect on hillslope form.

As well as bringing about increases in the effectiveness of particular processes, land-use changes may alter the balance of processes acting on particular hillslopes. In terms of hillslope form, the most obvious such change is a shift from wash processes to mass movements as the dominant mechanism of erosion. However, insofar as all processes lead to their own characteristic hillslope forms, any shift in the balance of processes can be expected to lead to some changes in hillslope form. At the present time it is easier to document the effects that past land-use changes have had than to predict those that future changes may bring about. Slaymaker (1982) has identified the problem of inadequate definition and quantification of land-use change to match more easily measured effects. Furthermore, it cannot always be assumed that changes in hillslope-process rates and mechanisms are wholly a function of land-use change. Moss and Walker (1978) have argued that a heavy vegetation cover may protect a hillslope from erosion long after the hillslope has ceased to be in equilibrium with prevailing conditions. Stripping of vegetation by man may precipitate the occurrence of a backlog of erosion that represents not just adjustment to new conditions but also to ones that prevailed before land-use change came about.

Figure 10.3: Gully formation. A possible long-term equilibrium response of the drainage network to land-use change

ARTIFICIAL LANDSCAPES

Requirements of those engaged in extractive industries to return the land to its former condition once quarrying and mining have ceased result in the need to create artificial landscapes (e.g. Figure 10.4). In these cases consideration

Figure 10.4: An artificial landscape. These hillslopes and the drainage line have been created from a reclaimed mining area

needs to be given not only to hillslope stability but to the overall appearance of the landscape. Ideally the artificial landscapes will be indistinguishable from the natural ones surrounding them. Such practice is in its infancy but it is a part of hillslope geomorphology in which digital elevation data could be particularly useful. Such data on natural landscapes could be used to plan the refashioning of spoil heaps to blend with surrounding landscapes.

HILLSLOPE FORM AND THE PREDICTION OF HAZARD

In using hillslopes as sites for the construction of buildings and roads, man transforms natural geomorphic processes into hazards. Of these processes, rapid mass movements, because of their suddenness and often spectacular effects, have been perceived as the main hazard. To what extent can this hazard be assessed? Although techniques for evaluating the factor of safety of potentially moving material are well established, it is difficult to apply these techniques to a whole hillslope. Natural variability of hillslope materials leads to uncertainty in estimates of the factor of safety (Ward, Li and Simons, 1981). Those who have attempted to define mass movement hazard for hillslopes have, therefore, adopted a probabilistic approach (Ingles, 1976; Ward *et al.*, 1981).

Ingles proposed delineating units of hillslopes according to their gradients and then assessing the probability of a mass movement on each unit according to its score over a number of criteria (Table 10.1). Ward *et al.* used less satisfactory grid cells as their basic mapping units. In this study the factor of safety of each grid cell was first assessed, followed by an estimate of the uncertainty with which the factor of safety was known for each grid cell. In the final stage these two pieces of information were combined to yield a hazard map.

In an assessment of the validity of the method used by Ward *et al.*, Simons, Li and Ward (1978) compared their hazard assessments with the actual distribution of landslides and obtained a better than 80 percent agreement between the actual distribution of landslides and the estimated hazardous cells. The existence of morphological information on hillslopes attesting

178

Table 10.1: Suggested probabilities to be assigned
to hillslope mapping units to indicate probability
of mass movements

DEMERIT PROBABILITIES (all for on or next to the site)		MERIT ALLOWANCES (all for on site only)	
Impeded drainage on or above a slope greater than 1 in 8	0.40	Good deep drainage	-0.40
Slope greater than 1 in 4	0.20	Slope less than 1 in 8	-0.20
Cleared land	0.15	Good tree cover	-0.15
Subsurface clay layer	0.05	Artificial toe loading (or head cuts)	-0.05
Inclined rock bedding	0.05	Horizontal rock bedding	.0.05
Artificial toe loading (or toe cuts)	0.05		
Ground water (springs)	0.05		
Surface cracking	0.03		
Seismic region 0.02			
Total	1.00		

Notes (a) Stabilization measures shall be designed sufficient to offset the degree of risk (demerit) for any site development
(b) Evidence of movements greater than 0.1 mm/yr in any given probability zone automatically increases the risk in that zone by 0.2
(c) If any of the above conditions are not known, allow maximum demerit
(d) Mapping units may be pooled, but not subdivided. If pooled the basis shall be the summation of area times probability divided by total area

Based on Ingles, (1976, p.3.6) Water Research
Foundation of Australia, Report No. 48.

to mass movement activity suggests that such
information might usefully be employed not merely
to test the validity of assessments but also to
identify areas of instability. Both grid cells
and hillslope units of uniform gradient fail to
make significant use of the information contained
within hillslope form pertinent to the assessment
of instability. Morphological mapping, as
employed by Brunsden and Jones (1972) allows a
greater input of this information. A combination
of morphological mapping with an evaluation of the
probability of mass movement within each
morphological unit may provide the best approach
to hazard assessment on hillslopes.

BIBLIOGRAPHY

Abrahams,A.D. (ed.) (1986) *Hillslope Processes*, Allen & Unwin, Boston, Mass.

----- and Parsons,A.J. (1977) 'Long Segments on Low-angled Hillslopes in New South Wales', *Area,9*, 124-127

----- and Parsons,A.J. (1987) 'Identification of Strength Equilibrium Rock Slopes: Further Statistical Considerations', *Earth Surface Processes and Landforms,12*, 631-635

-----, Parsons,A.J. and Hirsch,P.J. (1985) 'Hillslope Gradient - Particle Size Relations: Evidence for the Formation of Debris Slopes by Hydraulic Processes', *J.Geol.,93*, 347-357

-----, Parsons,A.J. and Luk,S.H.(1988) 'Hydrologic and Sediment Responses to Simulated Rainfall on Desert Hillslopes in Southern Arizona', *Catena* (in Press)

Ahnert,F. (1964) 'Quantitative Models of Slope Development as a Function of Waste Cover Thickness', paper presented at the meeting of the IGU Commission on Slope Evolution

----- (1970a) 'A Comparison of Theoretical Slope Models with Slopes in the Field', *Zeit. für Geomorph. Supplementband,9*, 88-101

----- (1970b) 'An Approach towards a Descriptive Classification of Slopes', *Zeit. für Geomorph. Supplementband,9*, 71-84

----- (1973) 'COSLOP 2 - A Comprehensive Model Program for Simulating Slope Profile Development', *Geocom Programs, 8*

----- (1976a) 'Investigations of Structural Influences on Surface Form by Theoretical Modelling', *Zeit. für Geomorph. Supplementband,24*, 11-22

----- (1976b) 'Brief Description of a Comprehensive Three-

Dimensional Process-Response Model of Landform Development', *Zeit. für Geomorph. Supplementband,25*, 29-49

Allen,J.R.L. (1969)   'The Maximum Slope Angle Attainable by Surfaces Underlain by Bulked Equal Spheroids with Variable Dimensional Ordering', *Geol. Soc. Am. Bull., 80*, 1923-1930

Anderle,R. (1987)   'Fractal Properies of Talus Slopes', Unpublished M.A. Dissertation, State University of New York at Buffalo

Anderson,E.W. (1972)   'Terracettes: A Suggested Classification', *Area,4*, 17-20

Anderson,M.G. and Calver,A. (1977) 'On the Persistence of Landscape Features Formed by a Large Flood', *Inst. Br. Geogr. Trans. New Series,2*, 243-254

Aniya,M. (1985) 'Contemporary Erosion Rate by Landsliding in Amahata River Basin, Japan, *Zeit. für Geomorph.,29*, 301-314

Armstrong,A.C.(1976)   'A Three-Dimensional Simulation of Slope Forms', *Zeit. für Geomorph. Supplementband,25*, 20-28

----- (1982) 'A Comment on the Continuity Equation Model of Slope Profile Development and its Boundary Conditions', *Earth Surface Processes and Landforms,7*, 283-284

Arnett,R.R. (1971)   'Slope Form and Geomorphological Processes: An Australian Example,' *Inst. Br. Geogr. Spec. Pubn.,3*, 81-92

Atkinson,T.C. (1971)   'Hydrology and Erosion in a Limestone Terrain', unpublished Ph.D. thesis, University of Bristol

Bakker,J.P. and Le Heux,J.W.N.(1946) 'Projective - Geometric Treatment of O.Lehmann's Theory of the Transformation of Steep Mountain Slopes', *K. Nederl. Akad. Wetens. Series B,49*, 533-547

----- and Le Heux,J.W.N. (1947) 'Theory on Central Rectilinear Recession of Slopes', *K. Nederl. Akad. Wetens. Series B,50*, 959-966 and 1154-1162

----- and Le Heux,J.W.N.(1950) 'Theory on Central Rectilinear Recession of Slopes',*K. Nederl. Akad. Wetens. Series B,53*, 1073-1084 and 1364-1374

----- and Le Heux,J.W.N. (1952)   'A Remarkable New Geomorphological Law', *K. Nederl. Akad. Wetens. Series B,55*, 399-410 and 554-571

Ballantyne,C.K. and Eckford,J.P. (1984) 'Characteristics and Evolution of Two Relict Talus Slopes in Scotland',

Scott. Geog. Mag.,100, 20-33

Band,L.E. (1985) 'Simulation of Slope Development and the Magnitude and Frequency of Overland Flow Erosion in an Abandoned Hydraulic Gold Mine' in M.J.Woldenberg, (ed.), *Models in Geomorphology*, Allen & Unwin, Boston, Mass. pp.191-211

Bass,N.W. (1929) 'The Geology of Cowley County, Kansas', *Kansas Geol. Survey Bull.,12*, 17-23

Beaty,C.B. (1962) 'The Effect of Moisture on Slope Stabilty: A Classic Example from Southern Alberta, Canada', *J. Geol.,80*, 362-366

Bell,S.A. (1985) 'Attributes of Drainage Basin Topography', unpublished Ph.D. Thesis, University of Durham.

Blench,T. (1951) *'Hydraulics of Sediment-Bearing Canals and Rivers'*, Evans Industries Ltd., Vancouver

Blong,R.J. (1972) 'Methods of Slope Profile Measurement in the Field', *Aust. Geog. Stud.,10*, 182-192

----- (1974) 'Landslide Form and Hillslope Morphometry: An Example from New Zealand', *Aust. Geogr.,12*, 425-438

----- (1975) 'Hillslope Morphometry and Classification: A New Zealand Example', *Zeit. für Geomorph.,19*, 405-429

Brawner,C.O. (1966) 'Slope Stability in Open Pit Mines', *Western Miner,39*, 56-72

Bridge,B.J. and Beckman,G.G. (1977) 'Slope Profiles of Cycloidal Form', *Science,198*, 610-612

Brunsden,D. and Jones,D.K.C. (1972) 'The Morphology of Degraded Landslide Slopes in Southwest Dorset', *Quart. J. Engng. Geol.,5*, 205-222

----- and Kesel,R.H. (1973) 'Slope Development on a Mississippi River Bluff in Historic Time', *J. Geol.,81*, 576-598

Bryan,K. (1925) 'The Papago Country, Arizona', *U.S.G.S. Water Supply Paper 499*

----- (1940) 'The Retreat of Slopes', *Annal. Ass. Am. Geogr.,30*, 254-268

Bryan,R.B. (1979) 'The Influence of Slope Angle on Soil Entrainment by Sheetwash and Rainsplash', *Earth Surface Processes,4*, 43-58

----- and Yair,A. (1982) 'Perspectives on Studies of Badland Geomorphology' in R.Bryan and A.Yair (eds.) *Badland Geomorphology and Piping*, Geobooks, Norwich, pp.1-12

Bucknam,R.C. and Anderson,R.E. (1979) 'Estimation of Fault-

Scarp Ages from a Scarp-Height - Slope-Angle Relationship', *Geology,7*, 11-14

Büdel,J. (1948) 'Das System der Klimatischen Geomorphologie (Beiträge zur Geomorphologie der Klimazonen und Vorzeitklimate V)', *Verhandlungen Deutscher Geographentag,27*, 65-100

----- (1953) 'Die "Periglazial" Morphologischen Wirkungen des Eiszeitklimas auf der ganzen Erde', *Erkunde,7*, 249-266

----- (1963) 'Klima-Genetische Geomorphologie', *Geographische Rundschau,15*, 269-285

----- (1977) 'Slope History and Slope Age', *Zeit. für Geomorph. Supplementband,28*, 14-29

Caine,N. (1969) 'A Model for Alpine Talus Slope Development by Slush Avalanching', *J. Geol.,77*, 92-100

Calef,W. and Newcomb,R. (1953) 'An Average Slope Map of Illinois', *Annal. Ass. Am. Geogr.,43*, 305-316

Campbell,I.A.(1970) 'Microrelief Measurements on Unvegetated Shale Slopes', *Prof. Geogr.,22*, 212-220

Carson,M.A. (1969) 'Models of Hillslope Development under Mass Failure', *Geog. Analysis,1*, 76-100

----- (1975) 'Threshold and Characteristic Angles of Straight Slopes' in E.Yatsu *et al.* (eds.), *Mass Wasting: 4th Guelph Symposium on Geomorphology*, pp.19-34

----- (1977) 'Angles of Repose, Angles of Shearing Resistance and Angles of Talus Slopes', *Earth Surface Processes, 2*, 363-380

----- and Kirkby,M.J. (1972) *Hillslope Form and Process*, Cambridge University Press

----- and Petley,D. (1970) 'The Existence of Threshold Hillslopes in the Denudation of the Landscape', *Inst. Br. Geogr. Trans.,49*, 71-96

Carter,C.A. and Chorley,R.J. (1961) 'Early Slope Development in an Expanding Stream System', *Geol. Mag.,98*, 117-130

Carter,G. and Nobes,M.J. (1980) 'The Application of Erosion Slowness Theory to Hillslope Formation', *Earth Surface Processes,5*, 131-141

Chandler,R.J. (1973) 'The Inclination of Talus, Arctic Talus Terraces, and other Slopes Composed of Granular Materials', *J. Geol.,81*, 1-14

Chapman,C.A.(1952) 'A new Quantitative Method of Topographic Analysis', *Am. J. Sci.,250*, 428-453

Bibliography

Chemekova, T, Yu and Chemekov,Yu F. (1975) 'Hillslope Terracettes', *Sov. Geog.: Review and Translation,16*,
609-615

Chorley,R.J. (1959) 'The Geomorphic Significance of some
Oxford Soils', *Am. J. Sci.,257*, 503-515

----- (1964a) 'The Nodal Position and Anomalous Character of
Slope Studies in Geomorphological Research', *Geog. J.*,
130, 503-506

----- (1964b) 'Geomorphological Evaluation of Factors Controlling Shearing Resistance of Surface Soils in
Sandstone', *J. Geophys. Res.,69*, 1507-1516

Christofoletti,A. and Tavares,A.C. (1976) 'Analise de Perfis
de Vertentes Esculpidas em Rochas do Grupo Nova Lima
(Quadrilatero Ferrifero, Minas Gerais)', *Noticia
Geomorfologica,16*, 41-56

Church,M. and Mark,D.M. (1980) 'On Size and Scale in Geomorphology', *Prog. Phys. Geog.,4*, 342-390

Churchill,R.R. (1979) 'A Field Technique for Profiling Precipitous Slopes', *Br. Geomorph. Res. Group Tech. Bull.*,
24, 29-34

----- (1981) 'Aspect-Related Differences in Badlands Slope
Morphology', *Annal. Ass. Am. Geogr.,71*, 374-388

----- (1982) 'Aspect-Induced Differences in Hillslope Processes', *Earth Surface Processes and Landforms,7*,
171-182

Clark,M.J. (1965) 'The Form of Chalk Slopes', *Southampton
Res. Series in Geog.,2*, 3-34

Cleaves,E.T., Godfrey,A.E. and Bricker,O.P. (1970) 'The
Geochemical Balance of a small Watershed and its
Geomorphic Implications', *Geol. Soc. Am. Bull.,81*,
3015-3032

Cole,K.L. and Mayer,L. (1982) 'Use of Packrat Middens to
Determine Rates of Cliff Retreat in the Eastern Grand
Canyon, Arizona', *Geology,10*, 597-599

Colman,S.M. and Watson,K. (1983) 'Ages Estimated from a
Diffusion Equation Model for Scarp Degradation',
Science,221, 263-265

Corbel,J. (1957) 'Les Karsts du nord-ouest de l'Europe',
Institute des Etudes Rhodaniennes, Lyons, Memoir,12

Coulomb,C.A. (1776) 'Essais sur une Application des Règles
des Maximis et Minimis à quelques Problems de Statique
Relatifs à l'Architecture', *Acad. Sci. Paris, Memoirs
Presentées par divers Savants*

184

Cox,N.J. (1975) 'Hillslope Profile Determination and Analysis' in A.Goudie, *et al.* (eds.), *Geomorphological Techniques*, pp.62-65

----- (1983) 'Dividing a Hillslope Profile into Components', Unpublished Manuscript

Crabtree,R.W. and Burt,T.P. (1983) 'Spatial Variation in Solutional Denudation and Soil Moisture over a Hillslope Hollow', *Earth Surface Processes and Landforms,8*, 151-160

Crampton,C.B. (1977) 'A Note on Asymmetric Valleys in the Central Mackenzie River Catchment, Canada', *Earth Surface Processes,2*, 427-429

Crowther,J. and Pitty,A.F. (1983) 'An Index of Microrelief Roughness, illustrated with Examples from Tropical Karst Terrain in West Malaysia', *Rev. de Geomorph. Dyn.,32*, 69-74

Crozier,M.J. (1973) 'Techniques for the Morphometric Analysis of Landslips', *Zeit. für Geomorph.,17*, 78-101

-----, Gage, M., Pettinga, J.R., Selby, M.J. and Wasson, R.J. (1982) 'The Stability of Hillslopes' in J.M.Soons, & M.J.Selby (eds.), *Landforms of New Zealand*, Longman, Aukland, pp.45-66

Culling,W.E.H. (1963) 'Soil Creep and the Development of Hillside Slopes,' *J. Geol.,71*, 127-161

----- and Dakto,M. (1987) 'The Fractal Geometry of the Soil-covered Landscape', *Earth Surface Processes and Landforms,12*, 369-385

Currey,D.R. (1964) 'A Preliminary Study of Valley Asymmetry in the Ogoturuk Creek Area, N.W. Alaska', *Arctic,17*, 84-98

Czudek,T. (1964) 'Periglacial Slope Development in the Area of the Bohemian Massif in Northern Moravia', *Biul. Peryglac.,14*, 169-193

Dalrymple,J.B., Blong,R.J. and Conacher,A.J. (1968) 'A Hypothetical Nine-unit Land-surface Model', *Zeit. für Geomorph., 12*, 60-76

Davis,W.M. (1898) 'The Grading of Mountain Slopes', *Science, 7*, 1449

----- (1899) 'The Geographical Cycle', *Geog. J.,14*, 481-504

----- (1900) 'Glacial Erosion in France, Switzerland and Norway', *Proc. Boston Soc. Natural Hist.,29*, 273-322

----- (1905) 'The Geographical Cycle in an Arid Climate', *J. Geol.,13*, 381-407

----- (1906) 'The Sculpture of Mountains by Glaciers, *Scott. Geog. Mag.,22*, 76-89

----- (1932) 'Piedmont Benchlands and Primärrumpfe', *Geol. Soc. Am. Bull.,43*, 399-440

Degraff,J.V. (1979) 'Initiation of Shallow Mass Movement by Vegetative-type Conversion', *Geology,7*, 426-429

Demirmen,F. (1975) 'Profile Analysis by Analytical Techniques: A New Approach', *Geog. Analysis,7*, 245-266

Dietrich,W.E. and Dorn,R. (1984) 'Significance of Thick Deposits of Colluvium on Hillslopes: A case study involving the use of Pollen Analysis in the Coastal Mountains of Northern California', *J. Geol.,92*, 147-158

Dietrich, W.E., Wilson, C.J. and Reneau, S.L. (1986) 'Hollows, Colluvium and Landslides in Soil-Mantled Landscapes' in A.D.Abrahams (ed.), *Hillslope Processes*, pp.361-388

Diseker,E.G. and Sheridan,J.M. (1971) 'Predicting Sediment Yield from Roadbanks', *Trans. ASAE,14*, 102-105

Dohrenwend,J.C. (1978) 'Systematic Valley Asymmetry in the Central California Coast Ranges', *Geol. Soc. Am. Bull., 89*, 891-900

Douglas G.R. (1980) 'Magnitude Frequency Study of Rockfall in Co. Antrim N. Ireland', *Earth Surface Processes,5*, 123-129

Dowling,J.W.F. (1968) 'Land Evaluation for Engineering Purposes in Northern Nigeria' in G.A.Stewart (ed.), *Land Evaluation*, Macmillan, Melbourne, pp.147-159

Dumanowski,B. (1960) 'Notes on the Evolution of Slopes in an Arid Climate', *Zeit. für Geomorph. Supplementband 1*, 178-189

Dunkerley,D.L. (1976) 'A Study of Long-term Slope Stability in the Sydney Basin, Australia', *Engng. Geol.,10*, 1-12

----- (1978) 'Climatic Geomorphology and Fully Developed Slopes: A discussion', *Catena,5*, 79-80

----- (1980) 'The Study of the Evolution of Slope Form over long periods of time: A review of Methodologies and some new Observational Data from Papua New Guinea', *Zeit. für Geomorph. 24*, 52-67

----- and Toy,T.J. (1978) 'Hillslope Form and Climate: Discussion and Reply', *Geol. Soc. Am. Bull.,89*, 1111-1114

Dunne,T. (1980) 'Formation and Control of Channel Networks', *Prog. Phys. Geog.,4*, 211-240

----- and Aubrey,B.F. (1986) 'Evaluation of Horton's Theory of Sheeetwash and Rill Erosion on the Basis of Field Experiments' in A.D.Abrahams (ed.), *Hillslope Processes*, pp.31-53

Dusseault,M.B. and Morgenstern,N.R. (1978) 'Characteristics of Natural Slopes in the Athabasca Oil Sands', *Can. Geotech. J.,15*, 202-215

Dwornik,S.E., Johnson,S.E., Little,J.E. and Walker,J.E. (1959) 'Microrelief-physiography-land use relationships', *Geol. Soc. Am. Bull.,70*, 1804

Dylik,J. (1968) 'The Significance of the Slope in Geomorphology', *Bull. Soc. Sci. Lett. de Lodz 19*, 1-19

East,T.J. and Gillieson,P.S. (1979) 'Scree Characteristics and Processes, Mt. Toowoonan, South-East Queensland, *Queensland Geog. J.,5*, 79-91

Ellison,W.D. (1947) 'Soil Erosion Studies', *Agric. Engng., 28*, 145-146, 197-201, 245-248, 297-300, 349-351, 402-405 and 442-444.

Emery,K.O. (1947) 'Asymmetric Valleys of San Diego County, California', *Bull. S. California Acad. Sci. Pt.2*, 61-71

Engelen,G. and Huybrechts,W. (1981) 'A Comparison of Manual and Automated Slope Maps', *Catena,8*, 239-249

Engelen,G.B. (1973) 'Runoff Processes and Slope Development in Badlands National Monument, South Dakota', *J. Hyd., 18*, 55-80

Everard,C.E. (1963) 'Contrasts in the Form and Evolution of Hill-side Slopes in Central Cyprus', *Inst. Br. Geogr. Trans.,32*, 31-47

----- (1964) 'Climate Change and Man as Factors in the Evolution of Slopes', *Geog. J.,130*, 498-502

Fair,T.J. (1947) 'Slope Form and Development in the Interior of Natal', *Geol. Soc. S. Africa Trans.,50*, 105-120

----- (1948a) 'Slope Form and Development in the Coastal Hinterland of Natal', *Geol. Soc. S. Africa Trans.,51*, 37-53

----- (1948b) 'Hillslopes and Pediments of the Semiarid Karoo', *S. African Geog. J.,30*, 71-79

Fisher,O. (1866) 'On the Disintegration of a Chalk Cliff', *Geol. Mag.,3*, 354-356

Ford,D.C. (1963) 'Aspects of the Geomorphology of the Mendip Hills', unpublished Ph.D. thesis, University of Bristol

Fourneau,R. (1960) 'Contribution a l'Etude des Versants Dans le Sud de la Moyenne Belgique et Dans le Nord de L'Entre

Sambre et Meuse. Influence de la Nature du Substratum', *Annal. Soc. Geol. Belg.,84*, 123-151

French,H.M. (1971) 'Slope Asymmetry of the Beaufort Plain, Northwest Banks Island, N.W.T., Canada, *Can. Jour. Earth Sci.,8*, 717-731

----- (1972) 'Asymmetrical Slope Development in the Chiltern Hills, *Biul. Peryglac.,21*, 51-74

Frostick,L.E. and Reid,I. (1982) 'Talluvial Processes, Mass Wasting and Slope Evolution in Arid Environments', *Zeit. für Geomorph. Supplementband,44*, 53-67

Frye,J.C. (1959) 'Climate and Lester King's "Uniformitarian Nature of Hillslopes"', *J. Geol.,67*, 111-113

Gagan,P. and Gunn,J. (1987) 'The Magnitude and Frequency of Rockfalls from Limestone Quarries', Paper presented at the BGRG Annual Conference, Oxford

Galloway,R.W. (1961) 'Solifluction in Scotland', *Scott. Geog. Mag.,77*, 75-87

Gardiner,V. and Dackombe,R.V. (1977) 'A Simple Method for Field Measurement of Slope Profiles' in *Shorter Technical Methods II, Br. Geomorph. Res. Group Tech. Bull.,18*, 9-18

Gardner,J.S. (1983) 'Rockfall Frequency and Distribution in the Highwood Pass Area, Canadian Rocky Mountains', *Zeit. für Geomorph.,27*, 311-324

Gerrard,A.J.W. and Robinson,D.A.(1971) 'Variability in Slope Measurements: A Discussion of the Effects of Different Recording Intervals and Micro-Relief in Slope Studies', *Inst. Br. Geogr. Trans.,54*, 45-54

Gilbert,G.K. (1877) *Report on the Geology of the Henry Mountains*, Washington

----- (1904) 'Systematic Asymmetry of Crest Lines in ·the High Sierra of California,' *J. Geol.,2*, 579-588

----- (1909) 'The Convexity of Hilltops', *J. Geol.,17*, 344-350

Gilg,A.W. (1973) 'A Note on Slope Measurement Techniques', *Area,5*, 114-117

Glock,W.S. (1932) 'Available Relief as a Factor of Control in the Profile of a Land Form', *J. Geol.,40*, 74-83

Gloriod,A. and Tricart,J. (1952) 'Etude Statistique de Vallees Asymetriques Sur la Feuille St. Pol au 1/50,000', *Rev. de Geomorph. Dyn.,3*, 88-98

Goodman,J.M. and Haigh,M.J. (1981) 'Slope Evolution on Abandoned Spoil Banks in Eastern Oklahoma', *Phys. Geog.,2*,

160-173

Goudie,A.S. and Bull,P.A. (1984) 'Slope Profile Change and Colluvial Deposition in Swaziland: An SEM Analysis', *Earth Surface Processes and Landforms,9,* 289-300

Gray,D.H. and Leiser,A.J. (1982) *Biotechnical Slope Protection and Erosion Control,* Van Nostrand Reinhold, New York.

Greenway,D.R. (1987) 'Vegetation and Slope Stability' in M.G. Anderson and K.S.Richards (eds.), *Slope Stability,* Wiley, Chichester, pp.187-230

Gregory,K.J. (1966) 'Aspect and Landforms in North East Yorkshire', *Biul. Peryglac.,15,* 115-120

----- and Brown,E.H. (1966) 'Data Processing and the Study of Land Form', *Zeit. für Geomorph.,10,* 237-263

Gretener,P.E. (1967) 'Significance of the Rare Event in Geology', *Am. Ass. Petrol. Geol. Bull.,55,* 2197-2206

Hack,J.T. and Goodlett,J.C. (1960) 'Geomorphology and Forest Ecology of a Mountain Region in the Central Appalachians', *U.S.G.S. Prof. Paper,347*

Hadley,R.F. (1961) 'Some Effects of Microclimate on Slope Morphology and Drainage Basin Development', *U.S.G.S. Prof. Paper 424B*

----- and Toy,T.J. (1977) 'Relation of Surficial Erosion on Hillslopes to Profile Geometry', *U.S.G.S. J. Res.,5,* 487-490

Haigh,M.J. (1977) 'The Retreat of Surface Mine Spoil Bank Slopes', *Prof. Geogr.,29,* 62-65

----- (1979) 'Ground Retreat and Slope Evolution on Plateau-type Colliery Spoil Mounds at Blaenavon, Gwent', *Inst. Br. Geogr. Trans. New Series,4,* 321-328

----- (1980) 'Slope Retreat and Gullying on Revegetated Surface Mine Dumps, Waun Hoslyn, Gwent', *Earth Surface Processes,5,* 77-79

----- (1985) 'Geomorphic Evolution of Oklahoma Roadcuts', *Zeit. für Geomorph.,29,* 439-452

----- and Rydout,G.B. (1987) 'Erosion Pin Measurement in a Desert Gully', in V.Gardiner (ed.), *International Geomorphology 1986 Part II,* Wiley, Chichester pp.419-436

Hembree,C.H. and Rainwater,F.H. (1961) 'Chemical Degradation on Opposite Flanks of the Wind River Range, Wyoming', *U.S.G.S. Water Supply Paper,1769*

Hencher,S.R. (1987) 'The Implications of Joints and Structures for Slope Stability', in M.G.Anderson and

K.S.Richards (eds.), *Slope Stability*, Wiley, Chichester, pp.145-186

Higgins,C.G. (1982) 'Grazing-Step Terracettes and their Significance', *Zeit. für Geomorph.*,*26*, 459-472

Hirano,M. (1966) 'A Study of a Mathematical Model of Slope Development', *Geog. Rev. Japan*,*39*, 606-617

----- (1968) 'A Mathematical Model of Slope Development', *J. Geosciences Osaka City University*,*11*, 13-52

----- (1972a) 'A Discussion of the Origin of Regularity in Variation of Valley-side Slopes', *Bull. Fac. Lit. Osaka City University*,*23*, 659-673

----- (1972b) 'Theory on Graded Slopes', *Geog. Rev. Japan*, *45*, 703-716

----- (1975) 'Simulation of Development Process of Interfluvial Slopes with Reference to Graded Form, *J. Geol.*,*83*, 113-123

----- (1976) 'Mathematical Model and the Concept of Equilibrium in Connection with Slope Shear Ratio,' *Zeit. für Geomorph. Supplementband 25*, 50-71

----- (1981) 'Intensity of Lateral Erosion by Rivers Deduced from the Asymmetry of Cross-Valley Profile', *Japan. Geomorph. Union Trans.*,*1-2*, 117-134

Hobson,R.D.(1972) 'Surface Roughness in Topography: Quantitative Approach' in R.J.Chorley (ed.), *Spatial Analysis in Geomorphology*, Methuen, London, pp.221-245

Hopkins,D.M. and Taber,B. (1962) 'Asymmetrical Valleys in Central Alaska' in *Abstracts for 1961*, *Geol. Soc. Am. Special Papers*,*68*, 116

Horton,C.W., Hoffmann,A.A.J. and Hempkins W.B. (1962) 'Mathematical Analysis of the Microstructure of an Area of the Bottom of Lake Travis', *Texas J. Sci.*,*14*, 131-141

Horton,R.E. (1945) 'Erosional Development of Streams and their Drainage Basins; Hydrophysical approach to Quantitative Morphology', *Geol. Soc. Am. Bull.*,*56*, 275-370

Hsu,K.J. (1983) 'Actualistic Catastrophism', *Sedimentology*, *30*, 3-9

Hupp,C.R. (1984) 'Dendrogeomorphic Evidence of Debris Flow Frequency and Magnitude at Mount Shasta, California', *Environ. Geol. and Water Sci.*,*6*, 121-128

Hutchinson,J.N. (1968) 'Mass Movement' in R.W.Fairbridge (ed.), *The Encyclopedia of Geomorphology*, New York, Reinhold, pp.688-695

Hutchinson,J.N. and Gostelow,T.P. (1976) 'The Development of an Abandoned Cliff in Cowdon Clay at Hadleigh, Essex' in A.W.Skempton and J.N.Hutchinson (eds.), *A Discussion of Valley Slopes and Cliffs in Southern England: Morphology, Mechanics and Quaternary History*, (Phil. Trans. Royal Soc. London A 283-1315), pp.557-604

Iida,T. and Okunishi,K. (1983) 'Development of Hillslopes due to Landslides', *Zeit. für Geomorph. Supplementband 46*, 67-77

Imeson,A.C. (1974) 'Solute Variations in Small Catchment Streams', *Inst. Br. Geogr. Spec. Pubn.,6*, 87-100

Ingles,O.G. (1976) 'Is There a Planning Process to Mitigate the Problem of Landslip?' in *Workshop Papers, Unstable Landforms, Water Research Foundation of Australia Report No. 48*, pp.3.1-3.8

Innes,J.L. (1983) 'Lichenometric Dating of Debris-flow Deposits in the Scottish Highlands', *Earth Surface Processes and Landforms,8*, 579-588

----- (1985) 'Magnitude-Frequency Relations of Debris Flows in Northwest Europe', *Geog. Annal. Series A,67A*, 23-32

Jahn,A. (1968) 'Morphological Slope Evolution by Linear and Surface Degradation', *Geog. Polonica,14*, 9-21

Jennings,J.N. (1985) *Karst Geomorphology*, Blackwell, Oxford

Jevons,W.S. (1887) *The Principles of Science*, Macmillan, London.

Johnson,N.M., Likens, G.E., Bormann, F.H. and Pierce, R.S. (1968) 'Rate of Chemical Weathering of Silicate Minerals in New Hampshire', *Geochimica et Cosmochimica Acta,33*, 455-481

Jones,J.A.A. (1982) 'Experimental Studies of Pipe Hydrology' in R.Bryan and A.Yair (eds.), *Badland Geomorphology and Piping*, Geobooks, Norwich, pp.355-370

----- (1987) 'The Effect of Soil Piping on Contributing Areas and Erosion Patterns', *Earth Surface Processes and Landforms,12*, 229-248

Kane,P.(1978) 'Origins of Valley Asymmetry at Sarah Canyon', *Yearbook,- Ass. Pacific Coast Geogr.,40*, 103-115

Kennedy,B.A. (1976) 'Valley-side Slopes and Climate' in E.Derbyshire (ed.), *Geomorphology and Climate*, Wiley, Chichester, pp.171-201

----- and Melton,M.A. (1972) 'Valley Asymmetry and Slope Forms of a Permafrost Area in Northwest Territories, Canada', *Inst. Br. Geogr. Special Pubn.,4*, 107-121

Kertesz,A. and Szilard,J. (1979) 'Some Problems of Slope Development Reflected in Slope-profile Investigations', *Geog. Polonica,41*, 21-26

King,L.C. (1953) 'Canons of Landscape Evolution', *Geol. Soc. Am. Bull.,64*, 721-752

----- (1957) 'The Uniformitarian Nature of Hillslopes', *Edinburgh Geol. Soc. Trans.,17*, 81-102

----- (1962) *Morphology of the Earth*, Oliver and Boyd, Edinburgh

Kirkby,M.J. (1967) 'Measurement and Theory of Soil Creep', *J. Geol.,75*, 359-378

----- (1969) 'Infiltration, Throughflow and Overland Flow' in R.J.Chorley (ed.), *Water, Earth and Man*, Methuen, London, pp.215-228

----- (1971) 'Hillslope Process-Response Models Based on the Continuity Equation', *Inst. Br. Geogr. Special Pubn.,3*,

----- (1976a) 'Hydrological Slope Models: The Influence of Climate' in E.Derbyshire (ed.), *Geomorphology and Climate*, Wiley, Chichester, pp.247-267

----- (1976b) 'Deterministic Continuous Slope Models', *Zeit. für Geomorph. Supplementband,25*, 1-19

----- (1978) 'Implications for Sediment Transport' in M.J.Kirkby (ed), *Hillslope Hydrology*, Wiley, Chichester, pp.325-363

----- (1983) 'The Continuity Equation Slope Model and Basal Boundary Conditions: A Further Comment', *Earth Surface Processes and Landforms,8*, 287-288

----- (1984) 'Modelling Cliff Development in South Wales: Savigear Re-viewed', *Zeit. für Geomorph.,28*, 405-426

----- (1985) 'A Model for the Evolution of Regolith Mantled Slopes' in M.J.Woldenberg (ed.), *Models in Geomorphology*, Allen & Unwin, Boston, Mass., pp.213-237

----- (1986) 'A Two-Dimensional Simulation Model for Slope and Stream Evolution' in A.D.Abrahams (ed.), *Hillslope Processes*, pp.203-222

----- and Chorley,R.J. (1967) 'Throughflow, Overland Flow and Erosion', *Bull. Interl. Ass. Scient. Hyd.,12*, 5-21

Kirkland,J.T. and Armstrong,J.C. (1982) 'Slope Movements Related to Expansive Soils on the Blackland Prairie, North Central Texas', in R.G.Craig and J.L.Craft (eds.) *Applied Geomorphology*, Allen and Unwin, Boston, Mass., pp.85-93

Klein,M. (1981) 'A Quantitative Approach to the Analysis of

Slope Roughness and Effective Slope Angle', *Catena,8*, 281-284

Knill,R.A. (1982) 'Lithologic Controls of Microrelief', Unpublished B.Sc. Dissertation, Department of Geography, University of Keele

Koons,D. (1955) 'Cliff Retreat in South-Western United States', *Am. J. Sci.,253*, 44-52

Kotarba,A. (1980) 'Climatically Controlled Asymmetry of Slopes in the Central Mongolian Uplands', *Bull. Acad. Polon. Sci., Serie des Sciences de la Terre,28*, 139-145

----- (1986) 'Granite Hillslope Morphology and Present-Day Processes in Semi-Arid Zone of Mongolia', *Geog. Polonica,52*, 125-133

Kumar,A. (1981) 'The Nature of Slope Profiles on some Residual Hills in the Jamalpur-Kiul Hills, Monghyr, India', *Zeit. für Geomorph.,25*, 391-399

Lake,P. (1928) 'On Hill Slopes', *Geol. Mag.,65*, 108-116

Lal,R. (1984) 'Effect of Slope Length on Erosion of some Alfisols in Western Nigeria', *Geoderma,33*, 181-189

Lambert,J.L.M. (1961) 'Contribution a l'Etude des Pentes du Condroz', *Annal. Soc. Geol. Belg.,84*, 241-250

Lawson,A.C. (1915) 'The Epigene Profiles of the Desert', *Univ. California Pubn. Bull. Dept. Geol.,9*, 23-48

Lehmann,O.(1933) 'Morphologische Theorie der Verwitterung von Steinschlagwänden', *Vierteljahrsschrift Naturforsch. Ges. Zurich, 78*, 83-126

Le Roux,J.S. (1976) 'The Average Slope of the Orange Free State', *S. African Geogr.,5*, 321-326

Lewis,W.V. (1939) 'Snow-Patch Erosion in Iceland', *Geog. J., 94*, 153-161

Liebling,R.S. and Scherp,H.S. (1983) 'Systematic Unequal Dissection of Opposing Valley Sides', *J. Glaciology,29*, 512-514

Lohnes,R.A. and Handy,R.L. (1968) 'Slope Angles in Friable Loess', *J. Geol.,76*, 247-258

Looman,H. (1956) 'Observations about some Differential Equations concerning Recession of Mountain Slopes I and II', *K. Nederl. Akad. Wetens., Series B*, 259-271 and 272-284

Luke,J.C. (1972) 'Mathematical Models for Landform Evolution', *J. Geophys. Res.,77*, 2460-2464

----- (1976) 'A Note on the use of Characteristics in Slope

193

Evolution Models', *Zeit.für Geomorph. Supplementband,25,* 114-119

Mabbutt,J.A. (1973) 'Geomorphology of Fowlers Gap Station' in J.A.Mabbutt (ed.), *Lands of Fowlers Gap Station, New South Wales, Fowlers Gap Arid Zone Research Station Research Series,3,* University of New South Wales, pp.85-121

----- (1977) *Desert Landforms,* Australian National University Press, Canberra

Madduma Bandara,C.M. (1974) 'The Orientation of Straight Slope Forms on the Hatton Plateau of Central Sri Lanka', *J. Tropical Geog.,38,* 37-44

Malaurie,J.N. (1952) 'Sur l'Asymetrie des Versants dans l'Ile de Disko, Groenland', *Acad. Sci. Paris, Comptes Rendus,234,* 1461-1462

Mandelbrot,B.B. (1975) 'Stochastic Models for the Earth's Relief, the Shape and the Fractal Dimension of the Coastlines, and the Number - Area Rule for Islands', *Proc. National Acad. Sci. U.S.A.,72,* 3825-3828

Mandelbrot,B.B. (1977) *Fractals; form, chance, and dimension,* Freeman, San Francisco

Marsh,W.M. and Koerner,J.M. (1972) 'Role of Moss in Slope Formation', *Ecology, 53,* 489-493

Mathier,L. and Roy,A.G. (1984) 'Variations de la Forme des Versants le Long d'un Cours d'Eau Miniature', *Geog. Phys. et Quaternaire,38,* 81-86

Matsukura,Y., Hayashida,S. and Maekado,A. (1984) 'Angles of Valley-side Slope made of "Shirasu" Ignimbrite in South Kyushu, Japan', *Zeit. für Geomorph.,28,* 179-191

Meade,R.H. (1982) 'Sources, Sinks and Storage of River Sediment in the Atlantic Drainage of the United States', *J. Geol.,90,* 235-252

Melton,M.A. (1957) 'An Analysis of the Relation among Elements of Climate, Surface Properties, and Geomorphology', *Tech. Report No. 11 ONR Project NR 389-422,* New York.

----- (1958) 'Geometric Properties of Mature Drainage Systems and their Representation in an E4 Phase Space', *J. Geol.,66,* 35-54

----- (1960) 'Intravalley Variation in Slope Angles Related to Microclimate and Erosional Environment', *Geol. Soc. Am. Bull.,71,* 133-144

Metcalf,J.R. (1966) 'Angle of Repose and Internal Friction', *Interl. J. Rock Mechanics and Mining Sci.,3,* 155-162

Meyer,L.D. and Kramer,L.A. (1968) 'Relation between Land-Slope Shape and Soil Erosion', Paper presented to the ASAE Chicago, 10-13 December

Meyer,L.D. and Monke,E.J. (1965) 'Mechanics of Soil Erosion by Rainfall and Overland Flow', *Trans. ASAE,8*, 572-577 and 580

Miller,J.P. (1961) 'Solutes in Small Streams draining Single Rock Types, Sangre de Cristo Range, New Mexico', *U.S.G.S. Water Supply Paper,1535-F*

Mills,H.H. (1978) 'Hillslope Evolution on the Pennington Formation, Central Tennessee: an Illustration of Dynamic Equilibrium', *J. Tennessee Acad. Sci.,53*, 150-153

----- (1981) 'Boulder Deposits and the Retreat of Mountain Slopes or "Gully Gravure" revisited', *J. Geol.,89*, 649-660

Molchanov,A.K. (1967) 'On the Study of Characteristic and Limiting Slope Angles in the Southern Regions of the Buryat A.S.S.R. (In Russian)', *Metody Geomorfologicheskikh Issledovannii,1*, 134-143 (English Translation RTS 5175 National Lending Library, Boston Spa, 1969)

Moon,B.P. (1977a) 'On the Validity of a Comprehensive Slope Development Model', *Zeit. für Geomorph.,21*, 401-410

----- (1977b) 'A Note on the Derivation of a Form Criterion for the Testing of Mathematical Slope Models', *S. African Geogr.,5*, 476-478

----- (1984) 'Refinement of a Technique for Determining Rock Mass Strength for Geomorphological Purposes', *Earth Surface Processes and Landforms,9*, 189-193

----- and Selby,M.J. (1983) 'Rock Mass Strength and Scarp Forms in Southern Africa', *Geog. Annal. Series A,65A*, 135-145

Mosley,M.P. (1973) 'Rainsplash and the Convexity of Badland Divides', *Zeit. für Geomorph. Supplementband,18*, 10-25

----- (1975) 'A Device for the Accurate Survey of Small Scale Slopes' in: *Shorter Technical Methods I, Br. Geomorph. Res. Group Tech. Bull.,17*, 3-6

----- and O'Loughlin,C. (1980) 'Slopes and Slope Processes', *Prog. Phys. Geog.,4*, 97-106

Moss,A.J. and Walker,P.H. (1978) 'Particle Transport by Continental Water Flow in Relation to Erosion, Deposition, Soils and Human Activity', *Sedimentary Geol.,20*, 81-139

Moss,R.P. (1965) 'Slope Development and Soil Morphology in

a part of South-West Nigeria', *J. Soil Sci.,16*, 192-209

Musgrave, G.W. (1947) 'Quantitative Evaluation of Factors in Water Erosion - A First Approximation', *J. Soil and Water Conservation,2*, 133-138

Nash,D. (1980a) 'Morphologic Dating of Degraded Normal Fault Scarps', *J. Geol.,88*, 353-360

----- (1980b) 'Forms of Bluffs Degraded for different lengths of time in Emmet County, Michigan, U.S.A.', *Earth Surface Processes,5*, 331-345

Newson,M.D. (1970) 'Studies of Chemical and Mechanical Erosion by Streams in Limestone Areas', unpublished Ph.D. thesis, University of Bristol

Nir,D. (1983) *Man, a Geomorphological Agent,* Keter Publishing House, Jerusalem

Oberlander,T.M. (1972) 'Morphogenesis of Granitic Boulder Slopes in the Mojave Desert, California,' *J. Geol.,80*, 1-20

----- (1974) 'Landscape Inheritance and the Pediment Problem in the Mojave Desert of Southern California', *Am. J. Sci.,274*, 849-875

Ohmori,H. (1979) 'A Statistical approach to Asymmetry in Roughness of Mountain Slopes in Japan from the viewpoint of Climatic Geomorphology', *Bull. Dept. Geog. University of Tokyo,11*, 77-92

Ollier,C.D. and Thomasson,A.J. (1957) 'Asymmetrical Valleys of the Chiltern Hills', *Geog. J.,123*, 71-80

Ollier,C.D. and Tuddenham,W.G. (1962) 'Slope Development at Coober Pedy, South Australia', *J. Geol. Soc. Australia, 9*, 91-105

O'Neill,M.P. and Mark,D.M. (1987) 'On the Frequency Distribution of Land Slope', *Earth Surface Processes and Landforms,12*, 127-136

Ongley,E.D. (1970) 'Determination of Rectilinear Profile Segments by Automatic Data Processing', *Zeit. für Geomorph.,14*, 383-391

Onstad,C.A., Larson,C.L., Wischmsmeier,L.F. and Young,R.A. (1967) 'A Method for Computing Soil Movement through a Field', *Trans. ASAE,10*, 747-749

Oxley,N.C. (1974) 'Suspended Sediment Delivery Rates and Solute Concentration of Stream Discharge in two Welsh Catchments', *Inst. Br. Geogr. Spec. Pubn.,6*, 141-153

Pain,C.F. (1986) 'Scarp Retreat and Slope Development near Picton, New South Wales, Australia', *Catena,13*, 227-239

Pallister,J.W. (1956a) 'Slope Form and Erosion Surfaces in Uganda', *Geol. Mag.,93*, 465-472

----- (1956b) 'Slope Development in Buganda', *Geog. J., 122*, 80-87

Papo,H.B. and Gelbman,E. (1984) 'Digital Terrain Models for Slopes and Curvatures', *Photogramm. Engng. and Remote Sensing,50*, 695-701

Parsons,A.J. (1973) 'A Stochastic Approach to the Description and Classification of Hillslopes', unpublished Ph.D. Thesis, University of Reading,

----- (1976a) 'A Markov Model for the Description and Classification of Hillslopes', *Math. Geol.,8*, 597-616

----- (1976b) 'An Example of the Application of Deductive Models to Field Measurement of Hillslope Form', *Zeit. für Geomorph. Supplementband,25*, 145-153

----- (1977) 'Curvature and Rectilinearity in Hillslope Profiles', *Area,9*, 246-251

----- (1978) 'A Technique for the Classification of Hillslope Forms', *Inst. Br. Geogr. Trans. New Series,3*, 432-443

----- (1979) 'Plan Form and Profile Form of Hillslopes', *Earth Surface Processes,4*, 395-402

----- (1982) 'Slope Profile Variability in First-Order Drainage Basins', *Earth Surface Processes and Landforms,7*, 71-78

----- (1987) 'Process, Form and Boundary Conditions along Valley-side Slopes' in V.Gardiner (ed.), *International Geomorphology 1986 Part II*, pp.89-104

----- and Abrahams A.D. (1987) 'Gradient-Particle Size Relations on Quartz Monzonite Debris Slopes in the Mojave Desert', *J. Geol.,95*, 423-432

Pearce,A.J. (1976a) 'Magnitude and Frequency of Erosion by Hortonian Overland Flow', *J. Geol.,84*, 65-80

----- (1976b) 'Geomorphic and Hydrologic Consequences of Vegetation Destruction, Sudbury, Ontario', *Can. J. Earth Sci.,13*, 1358-1373

-----, Black,R.D. and Nelson,C.S. (1981) 'Lithologic and Weathering Influences on Slope Form and Processes, Eastern Raukumara Range, New Zealand' in T.R.H.Davies and A.J.Pearce (eds.), *Erosion and Sediment Transport in Pacific Rim Steeplands, IAHS Publication, 132*, 95-122

Penck,W. (1924) *Die Morphologische Analyse, Ein Kapitel der Physikalischen Geologie*, Engelhorns, Stuttgart. English

197

translation with summaries, by H.Czech and K.C.Boswell, *Morphological Analysis of Landforms*, Macmillan, London, 1953.

Piexoto,J.P., Saltzman,B. and Teweles,S. (1964) 'Harmonic Analysis of the Topography along Parallels of the Earth', *J. Geophys. Res.,69*, 1501-1505

Pilgrim,D.H., Huff,D.D. and Steele,T.D. (1978) 'A Field Evaluation of Subsurface and Surface Runoff, II Runoff Processes', *J. Hyd. 38*, 319-341

Pitty,A.F. (1966) 'Some Problems in the Location and Delimitation of Slope Profiles', *Zeit. für Geomorph.,10*, 454-461

----- (1968a) 'A Simple Device for the Field Measurement of Hillslopes', *J. Geol.,76*, 717-720

----- (1968b) 'Some Comments on the Scope of Slope Analysis Based on Frequency Distributions', *Zeit. für Geomorph.,12*, 350-355

----- (1969) 'A Scheme for Hillslope Analysis. I. Initial Considerations and Calculations', *University of Hull Occasional Pubn. Geog.,9*

----- (1972) 'Statistical Comparison of Penckian and Davisian Views with Actual Slope Forms', *Rev. de Geomorph. Dyn.,21*, 171-176

Podmore,T.H. and Huggins,L.F. (1981) 'An Automated Profile Meter for Surface Roughness Measurements', *Trans. ASAE,24*, 663-665

Raisz,E. and Henry,J. (1937) 'An Average Slope Map of Southern New England', *Geog. Rev.,27*, 467-472

Rankine,W.J.M. (1857) 'On the Stability of Loose Earth', *Trans. Royal Soc. London,147*, 9-27

Rapp,A, (1959) 'Avalanches in Lappland - a description of little known forms of periglacial debris accumulation', *Geog. Annal.,41*, 34-48

----- (1960) 'Recent Development of Mountain Slopes in Kärkevagge and Surroundings, Northern Scandinavia', *Geog. Annal.,42*, 65-200

Rayner,J.N. (1972) 'The Application of Harmonic and Spectral Analysis to the Study of Terrain' in R.J.Chorley (ed.), *Spatial Analysis in Geomorphology*, Methuen, London, pp.283-302

Rice,R.M. and Pillsbury,N.H. (1982) 'Predicting Landslides in Clearcut Patches' in D.E.Walling (ed.), *Recent Developments in the Explanation and Prediction of Erosion and Sediment Yield, IAHS Pubn. 137*, 303-311

Rich,J.L. (1916) 'A Graphical Method of Determining the Average Inclination of a Land Surface from a Contour Map', *Indiana Acad. Sci. Trans.,9*, 195-199

Richards,K.S. (1977) 'Slope Form and Basal Stream Relationships: some Further Comments', *Earth Surface Processes,2* 87-95

Richardson,J.H. (1982) 'Some Implications of Tropical Forest Replacement in Jamaica', *Zeit. für Geomorph. Supplementband,44*, 107-118

Richter,E. (1901) 'Geomorphologische Untersuchungen in den Hochalpen', *Pett. Mitt. Erg.-Hefte, 132*

Robinson,M. and Blyth,K. (1982) 'The Effect of Forestry Drainage Operations on Upland Sediment Yields: A Case Study', *Earth Surface Processes and Landforms,7*, 85-90

Rouse,W.C. and Farhan,Y.I. (1976) 'Threshold Slopes in South Wales', *Quart. J. Engng. Geol.,9*, 327-338

Ruhe,R.V. (1975) 'Climatic Geomorphology and Fully Developed Slopes', *Catena,2*, 309-320

Saunders,I. and Young,A. (1983) 'Rates of Surface Processes on Slopes, Slope Retreat and Denudation', *Earth Surface Processes and Landforms,8*, 473-501

Savigear,R.A.G. (1952) 'Some Observations on Slope Development in South Wales', *Inst. Br. Geogr. Trans.,18*, 31-51

----- (1956) 'Technique and Terminology in the Investigation of Slope Forms', *Prem. Rapp. Comm. L'Etude Versants, Union Geog. Interl*, pp.66-75

----- (1960) 'Slopes and Hills in West Africa', *Zeit. für Geomorph. Supplementband,1*, 156-171

----- (1962) 'Some Observations on Slope Development in North Devon and North Cornwall', *Inst. Br. Geogr. Trans, 31*, 23-42

----- (1965) 'A Technique of Morphological Mapping', *Annal. Ass. Am. Geogr.,55*, 514-538

----- (1967) 'The Analysis and Classification of Slope Profile Forms', *Zeit. für Geomorph. Supplementband,9*, 271-290

Scheidegger,A.E. (1960) 'Analytical Theory of Slope Development by Undercutting', *J. Alberta Soc. Petrol. Geol.,8*, 202-206

----- (1961a) *Theoretical Geomorphology*, Springer, Berlin

----- (1961b) 'Mathematical Models of Slope Development', *Geol. Soc. Am. Bull.,72*, 37-50

199

----- (1964) 'Lithologic Variations in Slope Development Theory', *U.S.G.S. Circular,485*

Schloss,M. (1966) 'Quantifying Terrain Roughness on Lunar and Planetary Surfaces', *J. Spacecraft and Rockets,3,* 283-285

Schmidt,K.H. (1987) 'Erosional and Depositional Processes on Dryland Cuesta Scarps' paper presented at the Workshop on Erosion, Transport and Deposition Processes in Semi-arid and Arid Areas, Jerusalem

Schumm,S.A. (1973) 'Geomorphic Thresholds and Complex Response of Drainage Systems' in M.Morisawa (ed.), *Fluvial Geomorphology,* Binghampton, New York, pp.299-310

----- and Lichty,R.W. (1965) 'Time, Space and Causality in Geomorphology', *Am. J. Sci.,263,* 110-119

Selby,M.J. (1971) 'Slopes and their Development in an Ice-Free, Arid Area of Antarctica', *Geog. Annal. Series A,53A,* 235-245

----- (1974a) 'Dominant Geomorphic Events in Landform Evolution', *Bull. Interl. Ass. Engng. Geol.,9,* 85-89

----- (1974b) 'Slope Evolution in an Antarctic Oasis', *New Zealand Geogr.,30,* 18-34

----- (1980) 'A Rock Mass Strength Classification for Geomorphic Purposes: with Tests from Antarctica and New Zealand', *Zeit. für Geomorph.,24,* 31-51

----- (1982a) *Hillslope Materials and Processes,* Oxford University Press

----- (1982b) 'Controls on the Stability and Inclinations of Hillslopes Formed ôn Hard Rocks', *Earth Surface Processes and Landforms,7,* 449-467

Seret,G. (1963) 'Essai de Classification des Pentes en Famenne', *Zeit. für Geomorph.,7,* 71-85

Sharp,R.P. (1942) 'Soil Structures in the St. Elias Range, Yukon Territory', *J. Geomorph.,5,* 274-301

Sharpe,C.F.S. (1938) *Landslides and Related Phenomena,* Columbia, New York

Simanton,J.R., Rawitz,E. and Shirley,E.D. (1984) 'Effects of Rock Fragments on Erosion of Semiarid Rangeland Soils' in *Erosion and Productivity of Soils Containing Rock Fragments,* Soil Science Society of America, Madison, 65-72

Simonett,D.S. (1967) 'Landslide Distribution and Earthquakes in the Bewani and Torricelli Mountains, New Guinea' in J.N.Jennings and J.A.Mabbutt (eds.), *Landform Studies*

From Australia and New Guinea, Australian National University Press, Canberra, pp.64-84

Simons,D.B., Li,R.M. and Ward,T.J. (1978) 'Mapping of Potential Landslide Areas in terms of Slope Stability', Colorado State University Civil Engineering Report prepared for Rocky Mountain Forest and Range Experiment Station, Flagstaff, Arizona

Skempton,A.W. (1945) 'Earth Pressure and the Stability of Slopes' in *The Principles and Applications of Soil Mechanics*, Institution of Civil Engineers, London.

----- and De Lory,F.A. (1957) 'Stability of Natural Slopes in London Clay', *Proc. 4th Interl. Conference on Soil Mechanics and Foundation Engng.*, London,2, 378-381

Slaymaker,O. (1982) 'Land Use Effects on Sediment Yield and Quality', *Hydrobiologia,91*, 93-109

Smith,B.J. (1978) 'Aspect-Related Variations in Slope Angle near Beni Abbes, Western Algeria', *Geog. Annal. Series A,60A*, 175-180

----- (1981) 'Slope Evolution in the Gwari Hills, Kaduna State, Nigeria', *Singapore J. Tropical Geog.,2*, 57-67

Smith, D.D. and Wischmeier, W.H. (1957) 'Factors Affecting Sheet and Rill Erosion', *Am. Geophys. Union Trans.,38*, 889-896

Smith,T.R. and Bretherton,F.P. (1972) 'Stability and Conservation of Mass in Drainage Basin Evolution', *Water Resources Res.,8*, 1506-1529

Sneath,P.H.A. and Sokal,R.R. (1973) *Numerical Taxonomy*, Freeman, San Francisco

Soons,J.M. and Rainer,J.N. (1968) 'Micro-Climate and Erosion Processes in the Southern Alps, New Zealand', *Geog. Annal. Series A,50A*, 1-15

Souchez,R. (1963) 'Evolution des Versants et Theorie de la Plasticite', *Rev. Belg. de Geog.,87*, 10-94

----- (1966) 'Slow Mass Movement and Slope Evolution in Coherent and Homogeneous Rocks', *Bull. Soc. Belg. de Geol.,74*, 189-213

Speight,J.G. (1971) 'Log-Normality of Slope Distributions', *Zeit. für Geomorph.,15*, 290-311

Starkel,L. (1975) 'Characteristics and Evolution of the Asymmetrical Relief of the Sant Valley', *Bull. Acad. Polon. Sci. Serie des Sciences de la Terre,23*, 201-205

Statham,I. (1976) 'A Scree Slope Rockfall Model', *Earth Surface Processes,1*, 43-62

Bibliography

Sterr,H. (1985) 'Rates of Change and Degradation of Hill-slopes Formed in Unconsolidated Materials: a Morphometric Approach to date Quaternary Fault Scarps in Western Utah, U.S.A.', *Zeit. für Geomorph.,29*, 315-333

Stoddart,D.R. (1969) 'Climatic Geomorphology: Review and Re-Assessment', *Prog. Geog.,1*, 159-222

Stone,R.O. and Dugundji,J. (1965) 'A Study of Microrelief - Its Mapping, Classification and Quantification by Means of a Fourier Analysis', *Engng. Geol.,1*, 89-187

Strahler,A.N.(1950) 'Equilibrium Theory of Slopes Approached by Frequency Distribution Analysis', *Am. J. Sci.,248*, 673-696 and 800-814

----- (1956a) 'Quantitative Slope Analysis', *Geol. Soc. Am. Bull.,67*, 571-596

----- (1956b) 'The Nature of Induced Erosion and Aggradation', in W.L.Thomas (ed.), *Man's Role in Changing the Face of the Earth*, University of Chicago Press, pp.621-638

Summerfield,M.A. (1976) 'Slope Form and Basal Stream Relationships: A Case Study in the Western Basin of the Southern Pennines, England', *Earth Surface Processes,1*, 89-94

Sutherland,D.G., Ballantyne,C.K. and Walker,M.J.C. (1984) 'Late Quaternary Glaciation and Environmental Change of St. Kilda, Scotland', *Boreas,13*, 261-272

Tacconi,P., Billi,P. and Montani,C. (1982) 'Slope Length and Sediment Yield from Hilly Cropland' in D.E.Walling (ed.), *Recent Developments in the explanation and Prediction of Erosion and Sediment Yield, IAHS Publication,137*, 199-207

Terzaghi,K. (1962) 'Stability of Steep Slopes on Hard Unweathered Rocks', *Geotechnique,12*, 251-270

Thornes,J. (1973) 'Markov Chains and Slope Series: the Scale Problem', *Geog. Analysis,5*, 322-328

Thrower,N.J.W. and Cooke,R.U. (1968) 'Scales for Determining Slope from Topographic Maps', *Prof. Geogr.,20*, 181-186

Toy,T.J. (1977) 'Hillslope Form and Climate', *Geol. Soc. Am. Bull.,88*, 16-22

----- (1983) 'A Linear Erosion/Elevation Measuring Instrument (LEMI)', *Earth Surface Processes and Landforms,8*, 313-322

Tricart,J. and Cailleux,A. (1965) *Introduction a la Geomorphologie Climatique*, S.E.D.E.S., Paris

202

Tricart,J. and Muslin,J. (1951) 'L'Etude Statistique des Versants', *Rev. de Geomorph. Dyn.,2*, 173-182

Troeh,F.R. (1965) 'Landform Equations fitted to Contour Maps', *Am. J. Sci.,263*, 616-627

Trofimov,A.M. and Moskovkin,V.M. (1983) 'Mathematical Simulation of Stable and Equilibrium River Bed Profiles and Slopes', *Earth Surface Processes and Landforms,8*, 383-390

----- and Moskovkin,V.M. (1984) 'Diffusion Models of Slope Development', *Earth Surface Processes and Landforms,9*, 435-453

Trudgill,S.(1985) *Limestone Geomorphology*, Longman, London,

Tuck,R. (1935) 'Asymmetrical Topography in High Latitudes Resulting from Alpine Glacial Erosion', *J. Geol.,43*, 530-538

Turner, H. (1977) 'A Comparison of some Methods of Slope Measurement from Large Scale Air Photos', *Photogrammetria,32*, 209-238

Tylor,A, (1875) 'Action of Denuding Agencies', *Geol. Mag.*, *22*, 433-473

Van Burkalow,A. (1945) 'Angle of Repose and Angle of Sliding Friction: An Experimental Study', *Geol. Soc. Am. Bull.,56*, 669-707

Van Dijk,W. and Le Heux,J.W.N. (1952) 'Theory of Parallel Rectilinear Slope Recession I and II', *K. Nederl. Akad. Wetens. Proc. Series B,55*, 115-122 and 123-129

Varnes,D.J. (1958) 'Landslide Types and Processes', *Highway Research Board Special Report,29*, 20-47

Vincent,P.J. and Clarke,J.V. (1976) 'The Terracette Enigma - A Review', *Biul. Peryglac.,25*, 65-77

Walker,E.H. (1948) 'Differential Erosion on Slopes of Northern and Southern Exposure in Western Wyoming (Abs.)', *Geol. Soc. Am. Bull.,59*, 1360

Wallace,R.E.(1977) 'Profiles and Ages of Young Fault Scarps, North-Central Nevada', *Geol. Soc. Am. Bull.,88*, 1267-1281

Walling,D.E. (1971) 'Sediment Dynamics of Small Instrumented Catchments in South East Devon', *Devonshire Ass. Trans.*, *103*, 147-165

Ward,T.J., Li,R.M. and Simons,D.B. (1981) 'Use of a Mathematical Model for Estimating Potential Landslide Sites in Steep Forested Drainage Basins', in *T.R.H.Davies and A.J.Pearce (eds.) Erosion and Sediment Transport in*

Bibliography

Pacific Rim Steeplands, IAHS Publication, 132, 21-41

Waters,R.S. (1958)   'Morphological Mapping', *Geography,43,* 10-17

----- (1962) 'Altiplanation Terraces and Slope Development in Vest-Spitsbergen and South-West England', *Biul. Peryglac.,11,* 89-101

Waylen,M.J. (1976) 'Aspects of the Hydrochemistry of a Small Drainage Basin', unpublished Ph.D. thesis, University of Bristol

----- (1979) 'Chemical Weathering in a Drainage Basin Underlain by Old Red Sandstone', *Earth Surface Processes,4,* 167-178

Welch,D.M. (1970) 'Substitution of Space for Time in a Study of Slope Development', *J. Geol.,78* 234-239

Wentworth,C.K. (1930)   'A Simplified Method of Determining the Average Slope of Land Surfaces', *Am. J. Sci.,20,* 184-194

Whalley,W.B., Douglas,G.R. and Jonsson,A. (1983) 'The Magnitude and Frequency of Large Rockslides in Iceland in the Postglacial', *Geog. Annal. Series A,65A,* 99-109

White,J.F. (1966) 'Convex-Concave Landscapes - a Geometrical Study', *Ohio J. Sci.,66,* 592-608

Whitten,E.H.T. (1964) 'Process-Response Models in Geology', *Geol. Soc. Am. Bull.,75,* 455-464

Wilson, P.(1978)   'Simple Analysis of Scree Profile Data', *Amateur Geologist,8,* 32-39

Wirthmann,A.(1977) 'Erosional Slope Development in Different Climates', *Zeit. für Geomorph. Supplementband,28,* 42-61

Wolman,M.G. and Miller,J.P. (1960) 'Magnitude and Frequency of Forces in Geomorphic Processes', *J. Geol.,68,* 54-74

Wood,A. (1942) 'The Development of Hillside Slopes', *Proc. Geol. Ass.,53,* 128-140

Woods,E.B. (1974)   'Spatial Variation in Hillslope Profiles in the Cumberland Plateau, Kentucky', *Prof. Geogr.,26,* 416-420

Yair,A. (1973) 'Theoretical Considerations on the Evolution of Convex Hillslopes', *Zeit. für Geomorph. Supplementband,18,* 1-9

----- and Klein, M. (1973) 'The Influence of Surface Properties on Flow and Erosion Processes on Debris Covered Slopes in an Arid Area', *Catena,1,* 1-18

Young,A. (1961) 'Characteristic and Limiting Slope Angles',

Zeit. für Geomorph.,5, 126-131

----- (1963) 'Deductive Models of Slope Evolution', *Neue Beiträge Zur Internationalen Hangforschung*, Vandenhoeck and Ruprechht, Gottingen, pp.45-66

----- (1964) 'Slope Profile Analysis', *Zeit für Geomorph. Supplementband,5*, 17-27

----- (1970) 'Slope Form in the Xavantina-Cachimbo area', *Geog. J.,136*, 383-392

----- (1971) 'Slope Profile Analysis: the System of Best Units', *Inst. Br. Geog. Special Pubn.,3*, 1-13

----- (1972) *Slopes*, Oliver and Boyd, Edinburgh

----- (1974a) 'Slope Profile Survey', *Br. Geomorph. Res. Group Tech. Bull.,11*,

----- (1974b) 'The Rate of Slope Retreat' in E.H.Brown and R.Waters (eds.), *Progress in Geomorphology: Papers in Honour of David L. Linton (Inst. Br. Geogr. Special Pubn.,7)*, 65-78

----- (1978) 'Slopes: 1970-1975' in C.Embleton, D.Brunsden and D.K.C. Jones (eds.), *Geomorphology: Present Problems and Future Prospects*, pp.73-83

Young,R.A. and Mutchler,C.K. (1967) 'Effect of Slope Shape on Erosion and Runoff', *ASAE Paper No. 67-706*

Young,R.W. (1983) 'The Tempo of Geomorphological Change: Evidence from Southeastern Australia', *J. Geol.,91*, 221-230

Zeimer,R.R. (1981) 'Roots and the Stability of Forested Slopes', in T.R.H.Davies and A.J.Pearce (eds.) *Erosion and Sediment Transport in Pacific Rim Steeplands, IAHS Publication, 132*, 343-361

Zingg,A.W. (1940) 'Degree and Length of Land Slope as it Affects Soil Loss in Runoff', *Agric. Engng. 21*, 59-64

INDEX

effect on microrelief 56-7
effect on processes 48, 54-5
influence of lithology 50
climatic geomorphology 47
coefficient of friction 87
cohesion 89-90
components 25
  associations 41-2
  consociations 41-2
constant slope 142
continuity equation 109, 155, 157
convergence 162
creep 29; *see also* soil creep, heave
critical height (H_c) 69-70, 86, 89, 96

Davis's scheme for hillslope evolution 138-9, *Figure 8.2; see also* slope decline
debris flows 167-8, 175, *Figure 7.6*
debris slopes 79-80, 84, *Figure 5.9*
definitions 5, 6
dendrochronology 168
denudation rates
  see rates of denudation
description 23-39
desert gilgai 59
detachment 107-8
  continuum 107-8, *Figure 7.1*
  control 107
diffusivity models 153-4
digital terrain models 18, 20, 25, 178
discontinuities 10
dominant events 130
dominant processes 130-1
drainage basins 6, 102-6
drainage density 96
drainage divides 11
duricrusts 52-3, 84, *Figures 4.3, 4.4*

earthflow *Figure 7.5*
elements 31

identification 26
equifinality 162
equilibrium 169
erosion rates; *see* rates of denudation
ERS-1 19
evolution 1-5, 137-64
  classification of studies 137, *Figure 8.1*
  effect of climate 55-6
  effect of rock type 78, 84-6
  effect of vegetation 94
  evaluation of models 159-63
  in unconsolidated deposits 90-1, *Figure 5.14*
  mathematical models 148-54
  process-response models 155-58
  qualitative inferential models 144-8
  qualitative theories 138-44
  rates 165-6, 171
  timescales 165

facets 9-10, 20
factor of safety 119-20
fault scarps 96
Fisher's model of hillslope evolution 148-50, 152-3, *Figure 8.9*
footslopes 90-1
four-element hillslope model 1, 26, 50, 53, 75, 143, *Figure 4.1*
Fourier transforms 18, 32-3 36
fractal model 21
fractal surfaces 20, 38-9
free face 142
freeze-thaw 57, 63
frequency of processes 4

geographical cycle 138
glacial processes
  effects on valley asymmetry 63
graded waste sheet 138-9